350 Home Plans

ONE STORY DESIGNS UNDER 2000 Sq. Ft.

S0-BXS-260

HOME PLANNERS, INC.

23761 RESEARCH DRIVE
FARMINGTON HILLS, MICHIGAN 48024
TELEPHONE: (313) 477-1854

Contents

	Page
Index to Designs	3
How to Read Floor Plans and Blueprints	4
How to Choose a Contractor	5
How to Shop for Mortgage Money	6
Traditional One-Story Homes 1200 - 1599 Sq. Ft.	7
Optional Exteriors and Plans	29
Contemporary One-Story Homes 1200 - 1599 Sq. Ft.	47
Traditional One-Story Homes 1600 - 2000 Sq. Ft.	71
Contemporary One-Story Homes 1600 - 2000 Sq. Ft.	115
Homes for Restricted Budgets Under 1200 Sq. Ft.	153
One-Story Livability with Bonus Space Upstairs	183
One-Story Livability with Bonus Space Below	205
Vacation Homes for Leisure Living Lifestyles	217
One-Story Multi-Family Living	231
Home Planners' Services	241
The Plan Books	242
The Complete Blueprint Package	244
Before You Order	246
15 Perennial Favorites of Varying Types & Styles	247

Edited by: Net Gingras
Cover design by: D. M. Naidus

Published by Home Planners, Inc., 23761 Research Drive, Farmington Hills, Michigan 48024. All designs and illustrative material Copyright © MCMLXXXII by Home Planners, Inc. All rights reserved. Reproduction in any manner or form not permitted. Printed in the United States of America. International Standard Book Number (ISBN): 0-918894-28-X.

Index to Designs

DESIGN NO.	PAGE NO.	DESIGN NO.	PAGE NO.	DESIGN NO.	PAGE NO.	DESIGN NO.	PAGE NO.	DESIGN NO.	PAGE NO.
31000	67	31322	44	31867	150	32220	254	32704	78
31019	133	31323	37	31869	124	32232	110	32705	78
31021	67	31325	90	31884	145	32234	123	32706	78
31024	21	31326	76	31890	84	32236	251	32707	11
31025	20	31327	13	31891	120	32238	142	32718	195
31032	227	31330	76	31892	77	32261	98	32728	106
31033	66	31331	92	31896	83	32265	112	32737	88
31034	23	31336	215	31902	183	32272	210	32738	89
31036	148	31337	84	31904	194	32277	106	32741	116
31044	80	31342	48	31917	120	32278	192	32742	88
31057	63	31343	82	31919	81	32279	184	32743	117
31058	177	31346	109	31920	84	32283	252	32744	58
31065	48	31350	45	31922	56	32284	185	32753	58
31072	54	31351	45	31932	128	32286	184	32754	139
31073	60	31352	45	31938	36	32309	190	32755	153
31074	17	31357	48	31939	14	32310	61	32761	205
31075	18	31362	74	31943	62	32312	119	32769	207
31084	190	31364	154	31944	16	32316	90	32792	131
31088	28	31366	18	31945	46	32318	248	32795	138
31091	94	31367	22	31946	46	32327	112	32796	138
31094	23	31372	200	31947	129	32330	118	32797	130
31096	144	31373	176	31948	129	32332	137	32802	34
31100	82	31374	27	31949	109	32349	127	32803	34
31103	216	31379	181	31962	108	32351	134	32804	34
31104	186	31380	31	31967	197	32355	122	32805	35
31107	26	31381	31	31974	208	32356	249	32806	35
31113	182	31382	30	31976	213	32360	86	32807	35
31115	196	31383	30	31980	86	32363	135	32810	40
31126	142	31384	95	32016	232	32374	107	32811	40
31129	148	31385	95	32017	240	32380	126	32812	41
31130	94	31387	38	32018	231	32382	136	32813	41
31147	17	31388	38	32019	232	32383	150	32814	40
31153	146	31389	38	32021	234	32386	116	32815	40
31156	69	31390	140	32022	234	32395	186	32816	41
31173	12	31394	201	32023	234	32417	222	32817	41
31182	51	31395	180	32038	236	32426	218	32818	47
31186	102	31396	120	32039	232	32434	228	32821	52
31187	182	31399	160	32044	237	32458	220	32822	52
31188	24	31425	226	32106	64	32461	220	32824	43
31189	16	31453	221	32119	225	32479	229	32825	42
31190	14	31461	217	32122	24	32483	230	32828	238
31191	18	31463	222	32127	193	32504	211	32859	8
31193	12	31471	223	32128	256	32505	33	32869	239
31195	60	31477	218	32131	252	32510	201	33131	204
31197	21	31497	218	32136	250	32511	247	33139	146
31215	140	31522	166	32143	250	32520	252	33144	98
31216	172	31531	166	32145	188	32528	130	33163	66
31222	73	31706	68	32146	189	32533	104	33165	118
31237	122	31714	212	32153	162	32550	102	33167	152
31252	103	31726	111	32154	162	32557	114	33177	100
31254	64	31748	90	32156	182	32563	186	33181	142
31255	125	31754	254	32158	174	32565	32	33184	224
31272	74	31758	108	32159	159	32570	178	33189	200
31276	56	31759	125	32160	171	32583	206	33190	50
31279	178	31760	124	32161	171	32591	70	33195	172
31280	99	31765	152	32163	158	32593	10	33196	174
31281	172	31785	132	32164	159	32597	7	33197	190
31282	100	31790	198	32165	158	32603	100	33203	54
31283	150	31791	256	32166	158	32604	96	33204	20
31297	154	31793	199	32167	170	32605	96	33207	62
31298	209	31797	74	32168	170	32606	8	33208	176
31300	156	31798	132	32170	71	32607	179	33211	24
31301	157	31802	8	32171	248	32611	29	33212	54
31305	30	31803	26	32181	255	32612	29	33213	160
31307	31	31808	180	32182	65	32635	202	33214	181
31309	154	31812	214	32194	164	32636	203	33219	148
31311	166	31813	140	32195	165	32671	28	33221	156
31314	57	31815	22	32198	168	32672	72	33222	161
31316	14	31829	86	32199	169	32677	104	33223	157
31317	93	31862	80	32200	110	32678	105	33224	161
31319	50	31864	39	32201	147	32702	115	33226	175
31320	44	31865	39	32206	96	32703	58	33227	175
31321	44	31866	39	32210	113				

On the Cover: Cover Designs can be found on the following pages: Front cover - Design 32170, page 71. Back cover - top, Design 32550, page 102; middle, Design 32232, Page 110; bottom, Design 32330, page 118.

How to read floor plans and blueprints

Selecting the most suitable house plan for your family is a matter of matching your needs, tastes, and life-style against the many designs we offer. When you study the floor plans in this issue, and the blueprints that you may subsequently order, remember that they are simply a two-dimensional representation of what will eventually be a three-dimensional reality.

Floor plans are easy to read. Rooms are clearly labeled, with dimensions given in feet and inches. Most symbols are logical and self-explanatory: The location of bathroom fixtures, planters, fireplaces, tile floors, cabinets and counters, sinks, appliances, closets, sloped or beamed ceilings will be obvious.

A blueprint, although much more detailed, is also easy to read; all it demands is concentration. The blueprints that we offer come in many large sheets, each one of which contains a different kind of information. One sheet contains foundation and excavation drawings, another has a precise plot plan. An elevations sheet deals with the exterior walls of the house; section drawings show precise dimensions, fittings, doors, windows, and roof structures. Our detailed floor plans give the construction information needed by your contractor. And each set of blueprints contains a lengthy materials list with size and quantities of all necessary components. Using this list, a contractor and suppliers can make a start at calculating costs for you.

When you first study a floor plan or blueprint, imagine that you are walking through the house. By mentally visualizing each room in three dimensions, you can transform the technical data and symbols into something more real.

Start at the front door. It's preferable to have a foyer or entrance hall in which to receive guests. A closet here is desirable; a powder room is a plus.

Look for good traffic circulation as you study the floor plan. You should not have to pass all the way through one main room to reach another. From the entrance area you should have direct access to the three principal areas of a house—the living, work, and sleeping zones. For example, a foyer might provide separate entrances to the living room, kitchen, patio, and a hallway or staircase leading to the bedrooms.

Study the layout of each zone. Most people expect the living room to be protected from cross traffic. The kitchen, on the other hand, should connect with the dining room—and perhaps also the utility room, basement, garage, patio or deck, or a secondary entrance. A homemaker whose workday centers in the kitchen may have special requirements: a window that faces the backyard; a clear view of the family room where children play; a garage or driveway entrance that allows for a short trip with groceries; laundry facilities close at hand. Check for efficient placement of kitchen cabinets, counters, and appliances. Is there enough room in the kitchen for additional appliances, for eating in? Is there a dining nook?

Perhaps this part of the house contains a family room or a den/bedroom/office. It's advantageous to have a bathroom or powder room in this section.

As you study the plan, you may encounter a staircase, indicated by a group of parallel lines, the number of lines equaling the number of steps. Arrows labeled "up" mean that the staircase leads to a higher level, and those pointing down mean it leads to a lower one. Staircases in a split-level will have both up and down arrows on one staircase because two levels are depicted in one drawing and an extra level in another.

Notice the location of the stairways. Is too much floor space lost to them? Will you find yourself making too many trips?

Study the sleeping quarters. Are the bedrooms situated as you like? You may want the master bedroom near the kids, or you may want it as far away as possible. Is there at least one closet per person in each bedroom or a double one for a couple? Bathrooms should be convenient to each bedroom—if not adjoining, then with hallway access and on the same floor.

Once you are familiar with the relative positions of the rooms, look for such structural details as:

- Sufficient uninterrupted wall space for furniture arrangement.
- Adequate room dimensions.
- Potential heating or cooling problems—i.e., a room over a garage or next to the laundry.
- Window and door placement for good ventilation and natural light.
- Location of doorways—avoid having a basement staircase or a bathroom in view of the dining room.
- Adequate auxiliary space—closets, storage, bathrooms, countertops.
- Separation of activity areas. (Will noise from the recreation room disturb sleeping children or a parent at work?)

As you complete your mental walk through the house, bear in mind your family's long-range needs. A good house plan will allow for some adjustments now and additions in the future.

Each member of your family may find the listing of his, or her, favorite features a most helpful exercise. Why not try it?

How to choose a contractor

A contractor is part craftsman, part businessman, and part magician. As the person who will transform your dreams and drawings into a finished house, he will be responsible for the final cost of the structure, for the quality of the workmanship, and for the solving of all problems that occur quite naturally in the course of construction. Choose him as carefully as you would a business partner, because for the next several months that will be his role in your life.

As soon as you have a building site and house plans, start looking for a contractor, even if you do not plan to break ground for several months. Finding one suitable to build your house can take time, and once you have found him, you will have to be worked into his schedule. Those who are good are in demand and, where the season is short, they are often scheduling work up to a year in advance.

There are two types of residential contractors: the construction company and the carpenter-builder, often called a general contractor. Each of these has its advantages and disadvantages.

The carpenter-builder works directly on the job as the field foreman. Because his background is that of a craftsman, his workmanship is probably good—but his paperwork may be slow or sloppy. His overhead—which you pay for—is less than that of a large construction company. However, if the job drags on for any reason, his interest may flag because your project is overlapping his next job and eroding his profits.

Construction companies handle several projects concurrently. They have an office staff to keep the paperwork moving and an army of subcontractors they know they can count on. Though you can be confident that they will meet deadlines, they may sacrifice workmanship in order to do so. Because they emphasize efficiency, they are less personal to work with than a general contractor. Many will not work with an individual unless he is represented by an architect. The company and the architect speak the same language; it requires far more time to deal directly with a homeowner.

To find a reliable contractor, start by asking friends who have built homes for recommendations. Check with local lumber yards and building supply outlets for names of possible candidates.

Once you have several names in hand, ask the Chamber of Commerce, Better Business Bureau, or local department of consumer affairs for any information they might have on each of them. Keep in mind that these watchdog organizations can give only the number of complaints filed; they cannot tell you what percent of those claims were valid. Remember, too, that a large-volume operation is logically going to have more complaints against it than will an independent contractor.

Set up an interview with each of the potential candidates. Find out what his specialty is—custom houses, development houses, remodeling, or office buildings. Ask each to take you into—not just to the site of—houses he has built. Ask to see projects that are complete as well as work in progress, emphasizing that you are interested in projects comparable to yours. A $300,000 dentist's office will give you little insight into a contractor's craftsmanship.

Ask each contractor for bank references from both his commercial bank and any other lender he has worked with. If he is in good financial standing, he should have no qualms about giving you this information. Also ask if he offers a warranty on his work. Most will give you a one-year warranty on the structure; some offer as much as a ten-year warranty.

Ask for references, even though no contractor will give you the name of a dissatisfied customer. While previous clients may be pleased with a contractor's work overall, they may, for example, have had to wait three months after they moved in before they had any closet doors. Ask about his follow-through. Did he clean up the building site, or did the owner have to dispose of the refuse? Ask about his business organization. Did the paperwork go smoothly, or was there a delay in hooking up the sewer because he forgot to apply for a permit?

Talk to each of the candidates about fees. Most work on a "cost plus" basis; that is, the basic cost of the project—materials, subcontractors' services, wages of those working directly on the project, but not office help—plus his fee. Some have a fixed fee; others work on a percentage of the basic cost. A fixed fee is usually better for you if you can get one. If a contractor works on a percentage, ask for a cost breakdown of his best estimate and keep very careful track as the work progresses. A crafty contractor can always use a cost overrun to his advantage when working on a percentage.

Do not be overly suspicious of a contractor who won't work on a fixed fee. One who is very good and in great demand may not be willing to do so. He may also refuse to submit a competitive bid.

If the top two or three candidates are willing to submit competitive bids, give each a copy of the plans and your specifications for materials. If they are not each working from the same guidelines, the competitive bids will be of little value. Give each the same deadline for turning in a bid; two or three weeks is a reasonable period of time. If you are willing to go with the lowest bid, make an appointment with all of them and open the envelopes in front of them.

If one bid is remarkably low, the contractor may have made an honest error in his estimate. Do not try to hold him to it if he wants to withdraw his bid. Forcing him to build at too low a price could be disastrous for both you and him.

Though the above method sounds very fair and orderly, it is not always the best approach, especially if you are inexperienced. You may want to review the bids with your architect, if you have one, or with your lender to discuss which to accept. They may not recommend the lowest. A low bid does not necessarily mean that you will get quality with economy.

If the bids are relatively close, the most important consideration may not be money at all. How easily you can talk with a contractor and whether or not he inspires confidence are very important considerations. Any sign of a personality conflict between you and a contractor should be weighed when making a decision.

Once you have financing, you can sign a contract with the builder. Most have their own contract forms, but it is advisable to have a lawyer draw one up or, at the very least, review the standard contract. This usually costs a small flat fee.

A good contract should include the following:

- Plans and sketches of the work to be done, subject to your approval.
- A list of materials, including quantity, brand names, style or serial numbers. (Do not permit any "or equal" clause that will allow the contractor to make substitutions.)
- The terms—who (you or the lender) pays whom and when.
- A production schedule.
- The contractor's certification of insurance for workmen's compensation, damage, and liability.
- A rider stating that all changes, whether or not they increase the cost, must be submitted and approved in writing.

Of course, this list represents the least a contract should include. Once you have signed it, your plans are on the way to becoming a home.

A frequently asked question is: "Should I become my own general contractor?" Unless you have knowledge of construction, material purchasing, and experience supervising subcontractors, we do not recommend this route.

Most people who are in the market for a new home spend months searching for the right house plan and the ideal building site. Ironically, these same people often invest very little time shopping for the money to finance their new home, though the majority will have to live with the terms of their mortgage for as long as they live in the house.

The fact is that all banks are not alike, nor are the loans that they offer—and banks are not the only financial institutions that lend money for housing. The amount of down payment, interest rate, and period of the mortgage are all, to some extent, negotiable.

- Lending practices vary from one city and state to another. If you are a first-time builder or are new to an area, it is wise to hire a real estate (not divorce or general practice) attorney to help you unravel the maze of your specific area's laws, ordinances, and customs.
- Before talking with lenders, write down all your questions. Take notes during the conversation so you can make accurate comparisons.
- Do not be intimidated by financial officers. Keep in mind that *you are not begging for money,* you are buying it. Do not hesitate to reveal what other institutions are offering; they may be challenged to meet or better the terms.
- Use whatever clout you have. If you or your family have been banking with the same firm for years, let them know that they could lose your business if you can get a better deal elsewhere.
- Know your credit rights. The law prohibits lenders from considering only the husband's income when determining eligibility, a practice that previously kept many people out of the housing market. If you are turned down for a loan, you have a right to see a summary of the credit report and change any errors in it.

Some credit unions are now allowed to grant mortgages. A few insurance companies, pension funds, unions, and fraternal organizations also offer mortgage money to their

How to shop for mortgage money

membership, often at terms more favorable than those available in the commercial marketplace.

A GUIDE TO MORTGAGES

The types of mortgages available are far more various than most potential home buyers realize.

Traditional Loans

Conventional home loans have a fixed interest rate and fixed monthly payments. About 80 percent of the mortgage money in the United States is lent in this manner. Made by private lending institutions, these fixed rate loans are available to anyone whom the bank officials consider a good credit risk. The interest rate depends on the prevailing market for money and is slightly negotiable if you are willing to put down a large down payment. Most down payments range from 15 to 33 percent.

You can borrow as much money as the lender believes you can afford to pay off over the negotiated period of time—usually 20 to 30 years.

The FHA does not write loans; it insures them against default in order to encourage lenders to write loans for first-time buyers and people with limited incomes. The terms of these loans make them very attractive. The interest rate is fixed by FHA at 13½ percent, and you may be allowed to take as long as 25 to 30 years to pay it off.

The down payment also is substantially lower with an FHA-backed loan. At present it is set at 3 percent of the first $25,000 and 5 percent of the remainder, up to the $60,000 limit. This means that a loan on a $60,000 house would require a $750 down payment on the first $25,000 plus $1,750 on the remainder, for a total down payment of $2,500. In contrast, the down payment for the same house financed with a conventional loan could run as high as $20,000.

Anyone may apply for an FHA-insured loan, but both the borrower and the house must qualify.

The VA guarantees loans for eligible veterans, and the husbands and wives of those who died while in the service or from a service-related disability. The VA guarantees up to 60 percent of the loan or $27,500, whichever is less. Like the FHA, the VA determines the appraised value of the house, though with a VA loan, you can borrow any amount up to the appraised value.

The Farmers Home Administration offers the only loans made directly by the government. Families with limited incomes in rural areas can qualify if the house is in a community of less than 20,000 people and is outside of a large metropolitan area; if their income is less than $15,600; and if they can prove that they do not qualify for a conventional loan.

A GUIDE TO LENDERS

Where can you turn for home financing? Here is a list of sources for you to approach:

Savings and loan associations are the best place to start because they write well over half the mortgages in the United States on dwellings that house from one to four families. They generally offer favorable interest rates, require lower down payments, and allow more time to pay off loans than do other banks.

Savings banks, sometimes called mutual savings banks, are your next best bet. Like savings and loan associations, much of their business is concentrated in home mortgages.

Commercial banks write mortgages as a sideline, and when money is tight many will not write mortgages at all. They do hold about 15 percent of the mortgages in the country, however, and when the market is right, they can be very competitive.

Mortgage banking companies use the money of private investors to write home loans. They do a brisk business in government-backed loans, which other banks are reluctant to handle because of the time and paperwork required.

For more information, write Farmers Home Administration, Department of Agriculture, Washington, D.C. 20250, or contact your local office.

New loan instruments

If you think that the escalating cost of housing has squeezed you out of the market, take a look at the following new types of mortgages.

The graduated payment mortgage features a monthly obligation that gradually increases over a negotiated period of time—usually five to ten years. Though the payments begin lower, they stabilize at a higher monthly rate than a standard fixed rate mortgage. Little or no equity is built in the first years, a disadvantage if you decide to sell early in the mortgage period.

These loans are aimed at young people who can anticipate income increases that will enable them to meet the escalating payments. The size of the down payment is about the same or slightly higher than for a conventional loan, but you can qualify with a lower income. As of last year, savings and loan associations can write these loans, and the FHA now insures five different types.

The flexible loan insurance program (FLIP) requires that part of the down payment, which is about the same as a conventional loan, be placed in a pledged savings account. During the first five years of the mortgage, funds are drawn from this account to supplement the lower monthly payments.

The deferred interest mortgage, another graduated program, allows you to pay a lower rate of interest during the first few years and a higher rate in the later years of the mortgage. If the house is sold, the borrower must pay back all the interest, often with a prepayment penalty. Both the FLIP and deferred interest loans are very new and not yet widely available.

The variable rate mortgage is most widely available in California, but its popularity is growing. This instrument features a fluctuating interest rate that is linked to an economic indicator—usually the lender's cost of obtaining funds for lending. To protect the consumer against a sudden and disastrous increase, regulations limit the amount that the interest rate can increase over a given period of time.

To make these loans attractive, lenders offer them without prepayment penalties and with "assumption" clauses that allow another buyer to assume your mortgage should you sell.

Flexible payment mortgages allow young people who can anticipate rising incomes to enter the housing market sooner. They pay only the interest during the first few years; then the mortgage is amortized and the payments go up. This is a valuable option only for those people who intend to keep their home for several years because no equity is built in the lower payment period.

The reverse annuity mortgage is targeted for older people who have fixed incomes. This very new loan instrument allows those who qualify to tap into the equity on their houses. The lender pays them each month and collects the loan when the house is sold or the owner dies.

Traditional
One-Story Homes
1200 - 1599 Sq. Ft.

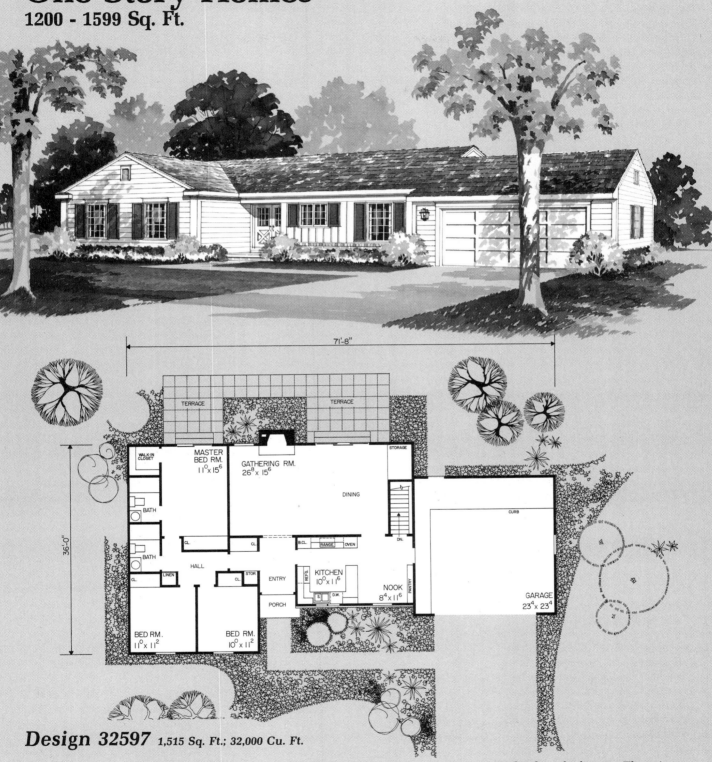

Design 32597 1,515 Sq. Ft.; 32,000 Cu. Ft.

● Whether it be a starter house you are after, or one in which to spend your retirement years, this pleasing frame home will provide a full measure of pride of ownership. The contrast of vertical and horizontal lines, the double front doors and the coach lamp post at the garage create an inviting exterior. The floor plan functions in an orderly and efficient manner. The 26 foot gathering room has a delightful view of the rear yard and will take care of those formal dining occasions. There are two full baths serving the three bedrooms. There is plenty of storage facilities, two sets of glass doors to the terraces, a fireplace in the gathering room, a basement and an attached two-car garage to act as a buffer against the wind. A delightful home, indeed.

Design 32859
1,599 Sq. Ft.; 37,497 Cu. Ft.

● Incorporated into the extremely popular basic one-story floor plan is a super-insulated structure. This means that it has double exterior walls separated by R-33 insulation and a raised roof truss that insures ceiling insulation will extend to the outer wall. More popularity is shown in the always popular Tudor facade. Enter the home through the air-locked vestibule to the foyer. To the left is the sleeping area. To the right of the foyer is the breakfast room, kitchen and stairs to the basement. Viewing the rear yard are the gathering and dining rooms. Study the technical details described in the blueprints of the wall section so you can better understand this super-insulated house.

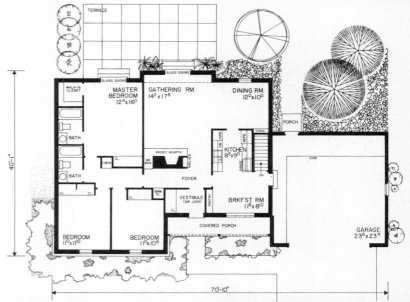

Design 31802
1,315 Sq. Ft.; 24,790 Cu. Ft.

● A small house which includes a full measure of big house livability features. The master bedroom has its extra wash room. In addition to the two bedrooms for the children, there is a study, or fourth bedroom. (This extra room offers the option to serve as a sewing, TV, music or even a guest room.) The living room is well situated and will not be bothered by cross-room traffic. The kitchen functions conveniently with the family-dining area. The stairs to the basement are just inside the entrance from the attached garage.

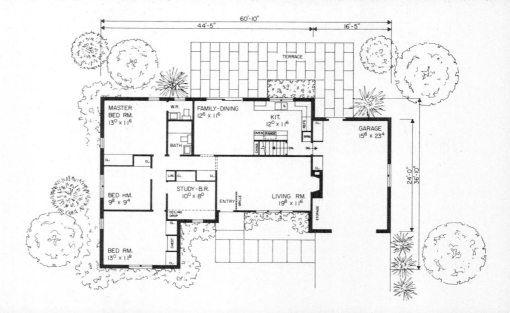

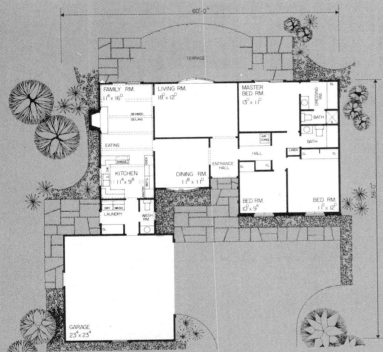

Design 32606
1,499 Sq. Ft.; 19,716 Cu. Ft.

● This modest sized house with its 1,499 square feet could hardly offer more in the way of exterior charm and interior livabiltiy. Measuring only 60 feet in width means it will not require a huge, expensive piece of property. The orientation of the garage and the front drive court are features which promote an economical use of property. In addition to the formal, separate living and dining rooms, there is the informal kitchen/family room area. Note the beamed ceiling, the fireplace, the sliding glass doors and the eating area of the family room.

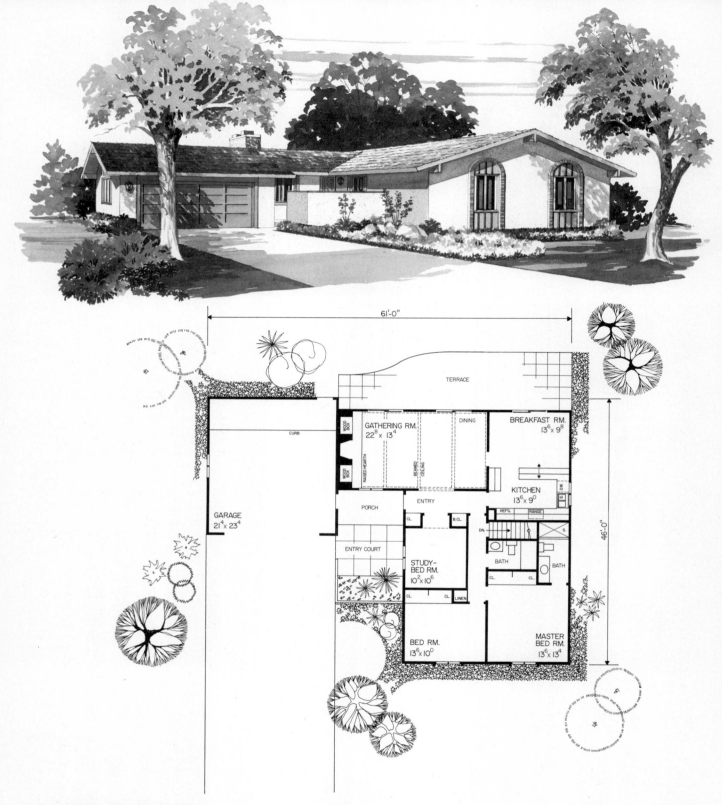

Design 32593 1,391 Sq. Ft.; 28,781 Cu. Ft.

● A fireplace wall! Including a raised hearth and two built-in wood boxes. A beamed ceiling, too. Inviting warmth in a spacious gathering room . . . more than 22' x 13' with ample space for a dining area. There's a sunny breakfast room, too, with sliding glass doors onto the terrace. And a pass-through from the kitchen. For efficiency, a U-shaped work area in the kitchen and lots of counter space. Two full baths. Three bedrooms! Including one with a private bath . . . and one suitable for use as a study, if that's your desire. This home is ideal for young families . . . equally perfect for those whose children are grown! It offers many of the attractive extras usually reserved for larger and more expensive designs. The appealing exterior will be appreciated.

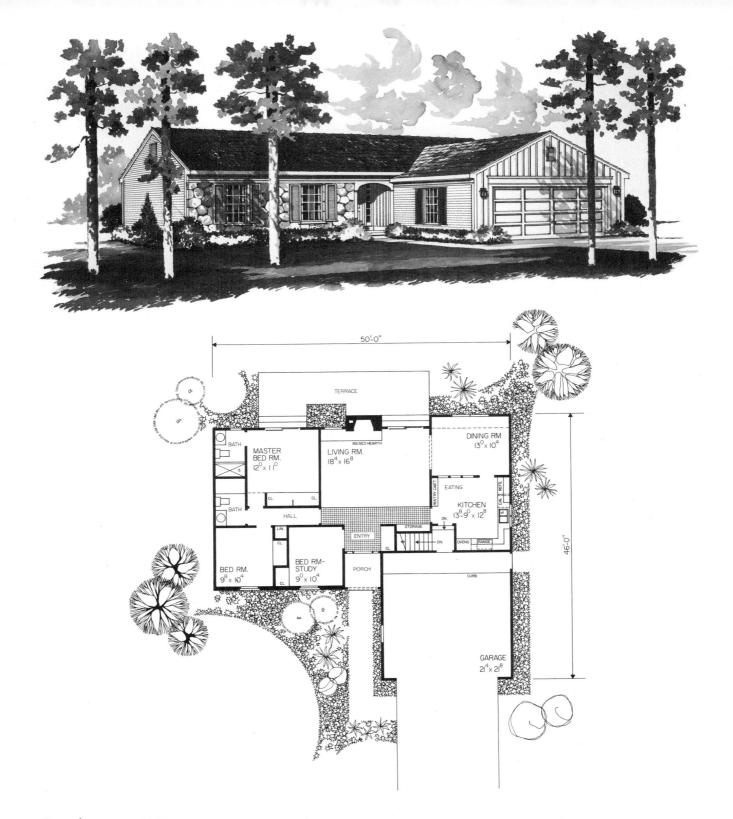

Design 32707   1,267 Sq. Ft.; 27,125 Cu. Ft.

● Here is a charming Early American adaptation that will serve as a picturesque and practical retirement home. Also, it will serve admirable those with a small family in search of an efficient, economically built home. The living area, highlighted by the raised hearth fireplace, is spacious. The kitchen features eating space and easy access to the garage and basement. The dining room is adjacent to the kitchen and views the rear yard. Then, there is the basement for recreation and hobby pursuits. The bedroom wing offers three bedrooms and two full baths. Don't miss the sliding doors to the terrace from the living room and the master bedroom. The storage units are plentiful including a pantry cabinet in the eating area of the kitchen.

Design 31173
1,500 Sq. Ft.; 18,230 Cu. Ft.

● A picturesque, L-shaped traditional one-story with an attractive covered porch. This floor plan is wonderfully zoned. All the major areas are but a few steps from the front entry hall. The living room with its bay window enables one to have a delightful view of the rear yard and also will lend itself to fine furniture arrangements and be free of cross-room traffic. Three bedrooms in the sleeping area plus a full bath. The homemaker will love the efficient kitchen as well as the separate laundry area. The extra wash room through the laundry is handy from the outdoors. The favorite spot will be the large family room. This multi-purpose area functions well with the kitchen. Built-in china cabinets are located in the two front corners of this area. Certainly a delightful feature. Two storage areas are in the garage. Note that the garage can be built to accommodate either one or two cars.

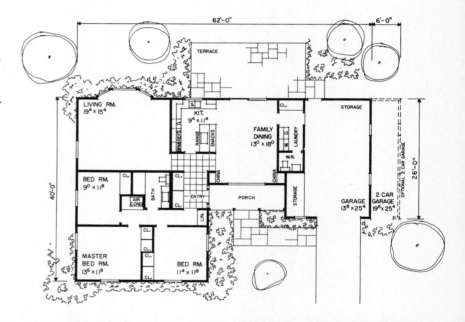

Design 31193
1,396 Sq. Ft.; 17,213 Cu. Ft.

● This L-shaped, one-story with its attached two-car garage incorporates many of the time-tested features of older New England. The attractive cut-up windows, the shutters, the panelled door, the fence and the wood siding contribute to the charm of the exterior. The floor plan is outstanding by virtue of such proven features as the separate dining room, the family-kitchen, the quiet living room with its bay window and the privacy of the bedroom area. The U-shaped work center will be easy to work in.

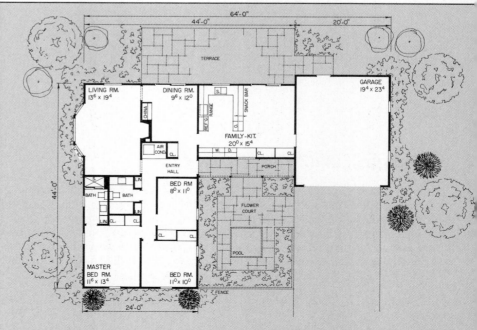

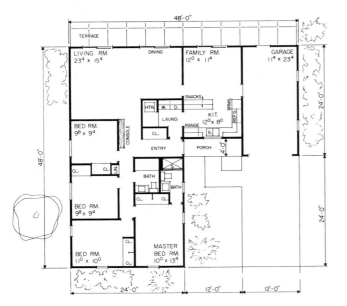

Design 31327
1,392 Sq. Ft.; 18,480 Cu. Ft.

● A design to overcome the restrictions of a relatively narrow building site. This home has one of the best one-story plans you'll ever see. Its four bedrooms and two baths form what amounts to a separate wing. The fourth bedroom is located next to the living room so it may serve as a study, if needed. The traffic plan pivoting in the front entry, is excellent. And the kitchen is in just the right place - next to the dining and family rooms, close to the front and side doors and near the laundry. Another fine quality of the kitchen is its snack bar pass-thru to the family room. Formal living will take place in the rear living/dining room. This area can open up to the rear terrace by two sets of sliding glass doors (another set is in the family room). An optional basement plan is included with the purchase of this design.

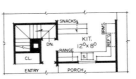

OPTIONAL BASEMENT PLAN

13

Design 31316
1,488 Sq. Ft.; 29,034 Cu. Ft.

● As picturesque as they come. Within 1,488 square feet, there are the fine sleeping facilities, two full baths, family room, a formal dining area, a big living room, an excellent kitchen and plenty of closets. Other noteworthy features include the fireplace, sliding glass doors to the rear terrace, a fine basement, a pass-thru to family room from kitchen, an attached garage and an appealing front porch.

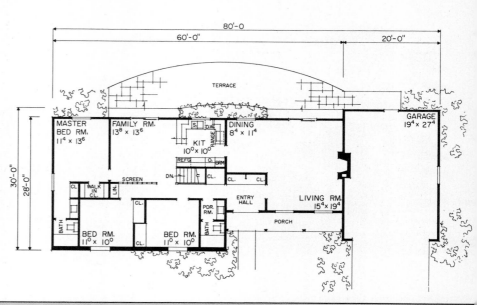

Design 31939
1,387 Sq. Ft.; 28,000 Cu. Ft.

● A finely proportioned house with more than its full share of charm. The brick veneer exterior contrasts pleasingly with the narrow horizontal siding of the oversized attached two-car garage. Perhaps the focal point of the exterior is the recessed front entrance with its double Colonial styled doors. The secondary service entrance through the garage to the kitchen area is a handy feature. Study the plan. It features three bedrooms, two full baths, living room with fireplace, front kitchen with an eating area, formal dining room, plenty of storage potential plus a basement for additional storage or perhaps to be developed as a recreational area.

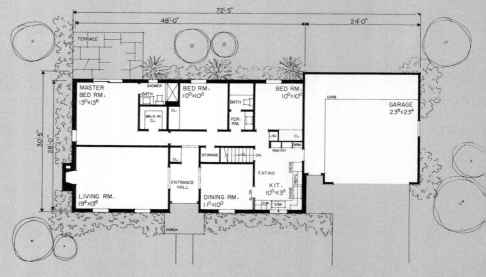

Design 31190
1,232 Sq. Ft.; 15,301 Cu. Ft.

● Build this L-shaped house on your lot with the garage door facing the street, or with the front entry door facing the street. Whichever orientation you choose, and you should take into consideration lot restriction and any particular view you might want to enjoy from the living areas, you will find the floor plan a very fine one. The center entrance hall features twin coat closets, a slate floor and an attractive six foot high room divider. This built-in unit has a planter on top with storage space below which is accessible from the living room.

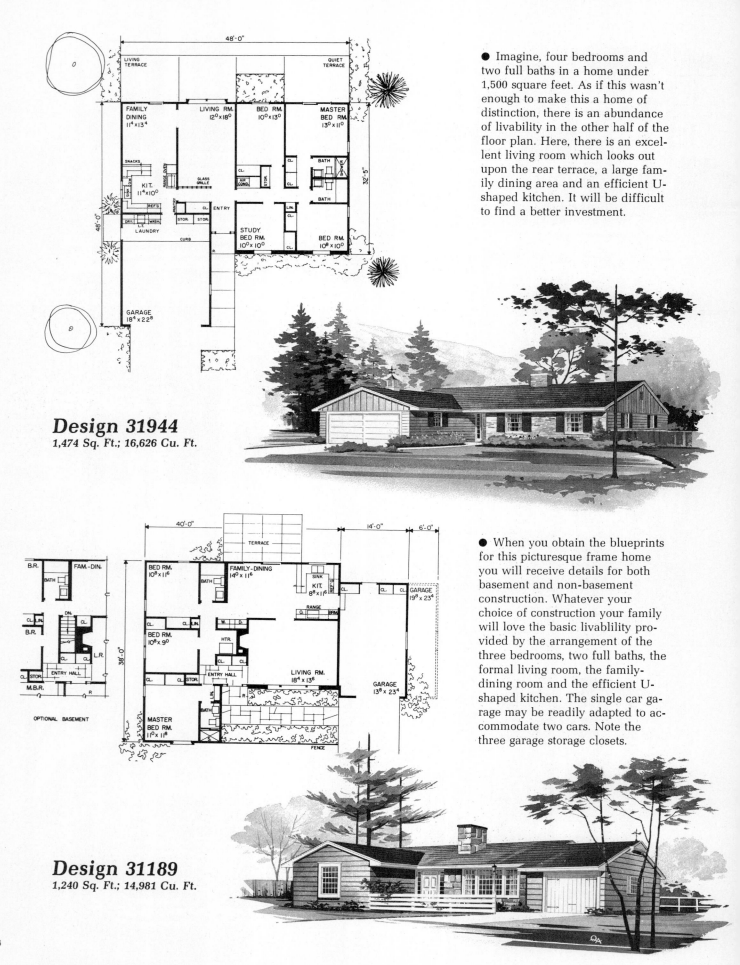

Design 31944
1,474 Sq. Ft.; 16,626 Cu. Ft.

● Imagine, four bedrooms and two full baths in a home under 1,500 square feet. As if this wasn't enough to make this a home of distinction, there is an abundance of livability in the other half of the floor plan. Here, there is an excellent living room which looks out upon the rear terrace, a large family dining area and an efficient U-shaped kitchen. It will be difficult to find a better investment.

Design 31189
1,240 Sq. Ft.; 14,981 Cu. Ft.

● When you obtain the blueprints for this picturesque frame home you will receive details for both basement and non-basement construction. Whatever your choice of construction your family will love the basic livablility provided by the arrangement of the three bedrooms, two full baths, the formal living room, the family-dining room and the efficient U-shaped kitchen. The single car garage may be readily adapted to accommodate two cars. Note the three garage storage closets.

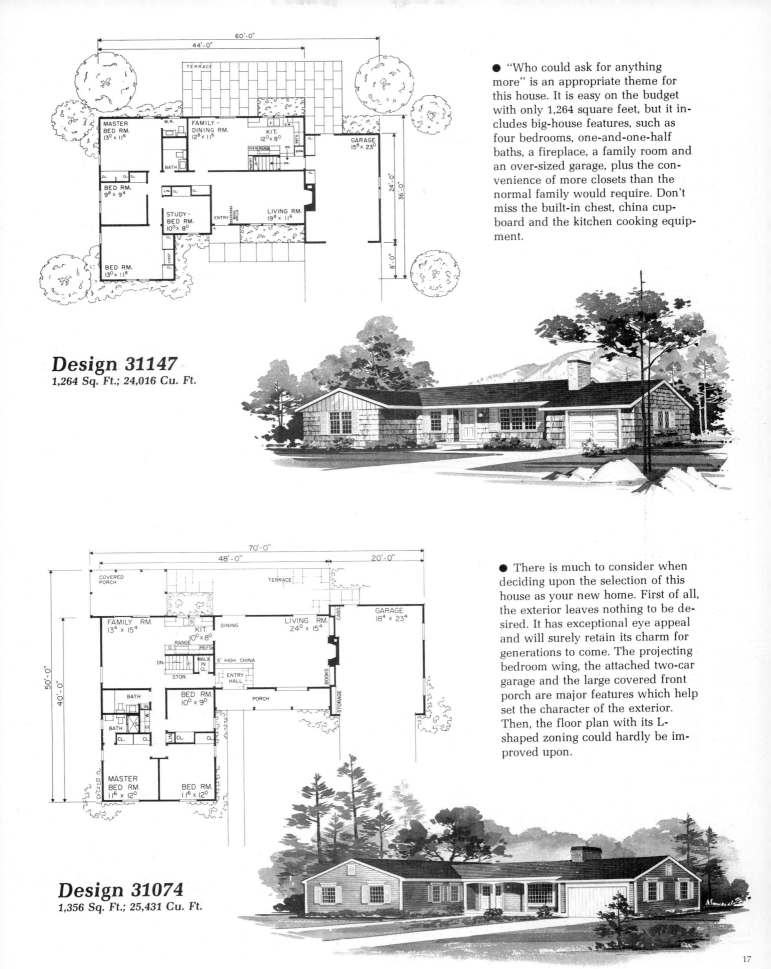

Design 31147
1,264 Sq. Ft.; 24,016 Cu. Ft.

● "Who could ask for anything more" is an appropriate theme for this house. It is easy on the budget with only 1,264 square feet, but it includes big-house features, such as four bedrooms, one-and-one-half baths, a fireplace, a family room and an over-sized garage, plus the convenience of more closets than the normal family would require. Don't miss the built-in chest, china cupboard and the kitchen cooking equipment.

Design 31074
1,356 Sq. Ft.; 25,431 Cu. Ft.

● There is much to consider when deciding upon the selection of this house as your new home. First of all, the exterior leaves nothing to be desired. It has exceptional eye appeal and will surely retain its charm for generations to come. The projecting bedroom wing, the attached two-car garage and the large covered front porch are major features which help set the character of the exterior. Then, the floor plan with its L-shaped zoning could hardly be improved upon.

Design 31075

1,232 Sq. Ft.; 24,123 Cu. Ft.

● This picturesque traditional one-story home has much to offer the young family. Because of its rectangular shape and its predominantly frame exterior, construction costs will be economical. Passing through the front entrance, visitors will be surprised to find so much livability in only 1,232 square feet. Consider these features: spacious formal living and dining area; two full baths; efficient kitchen; and large, rear family room. In addition there is the full basement for further recreational facilities and bulk storage. The attached garage is extra long to accommodate the storage of garden equipment, lawn furniture, bicycles, etc.

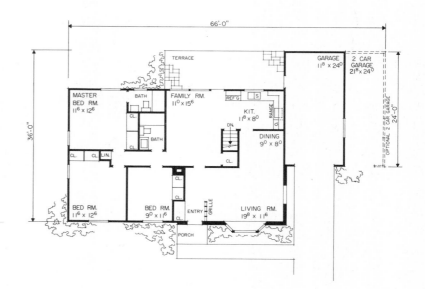

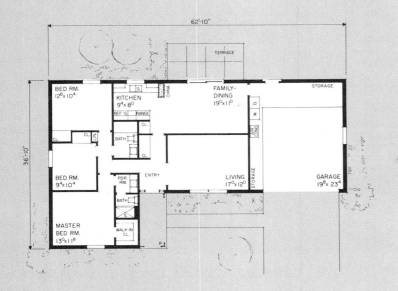

Design 31366
1,280 Sq. Ft.; 14,848 Cu. Ft.

● The extension of the main roof, along with the use of ornamental iron, vertical siding and glass side lites flanking the paneled door, all contribute to a delightful and inviting front entrance to this L-shaped design. There is much to recommend this design—from the attached two-car garage to the walk-in closet of the master bedroom. Don't overlook the compartmented master bath with its stall shower and powder room; the built-in china cabinet with an attractive planter above or the two closets right in the center of the house.

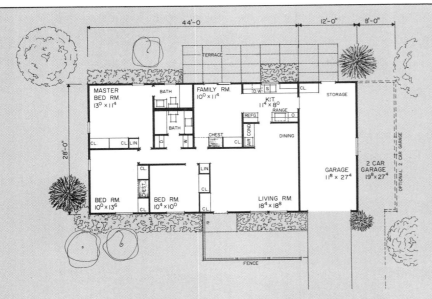

Design 31191
1,232 Sq. Ft.; 15,400 Cu. Ft.

● A careful study of the floor plan for this cozy appearing traditional home reveals a fine combination of features which add tremendously to convenient living. For instance, observe the wardrobe and storage facilities of the bedroom area. A built-in chest in the one bedroom and also one in the family room. Then, notice the economical plumbing of the two full back-to-back baths. Postively a great money saving feature for today and in the future. Further, don't overlook the location of the washer and dryer which have cupboards above the units themselves. Observe storage facilities. Optional two-car garage is available if necessary.

Design 31025   1,426 Sq. Ft.; 28,277 Cu. Ft.

● A real charmer. At the front of the house, parents have a private master suite with a full bath and generous closet space. Opposite the master bedroom is a secluded living room without any through-traffic, and a center fireplace that provides a focus of interest for arranging furniture. Recreational facilities may be developed in the full basement. This area also lends itself to developing additional storage.

● This traditional one-story home with its attached two-car garage virtually overflows with livability. There is not a foot of wasted space incorporated in this plan. Seldom is such good use made of a nominal amount of space. This home can be called upon to function as either a three or a four bedroom home. If the former, then the extra room may be utilized as a family room or study. There are two eating areas.

Design 33204   1,250 Sq. Ft.; 24,475 Cu. Ft.

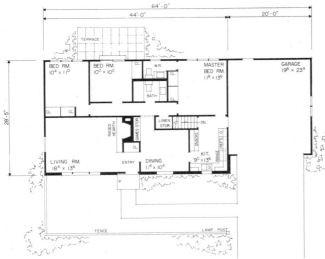

Design 31024   1,252 Sq. Ft.; 23,925 Cu. Ft.

● "Charming", is but one of the many words that could be chosen to describe this traditional home. While essentially a frame house, the front exterior features natural quarried stone. Below the overhanging roof, the windows and door treatment is most pleasing. The board fence with its lamp post completes a delightful picture. Highlighting the interior is the living room with its raised hearth fireplace.

● With the kitchen in this strategic location, the homemaker does not have to go through half of the house in order to get to the front door. The second bath with its stall shower is convenient to the third bedroom, the family area and the outdoor terrace. The main bath has a delightful built-in vanity. And how about the living room which has a large picture window and an attractive fireplace? Note storage closet.

Design 31197   1,232 Sq. Ft.; 24,123 Cu. Ft.

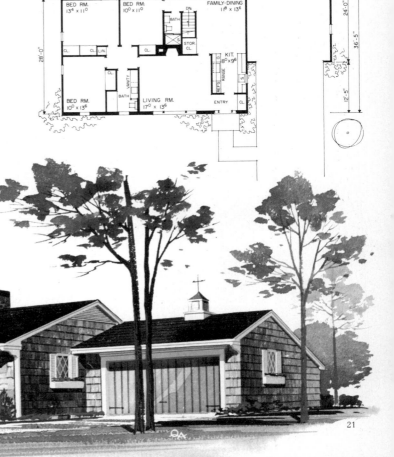

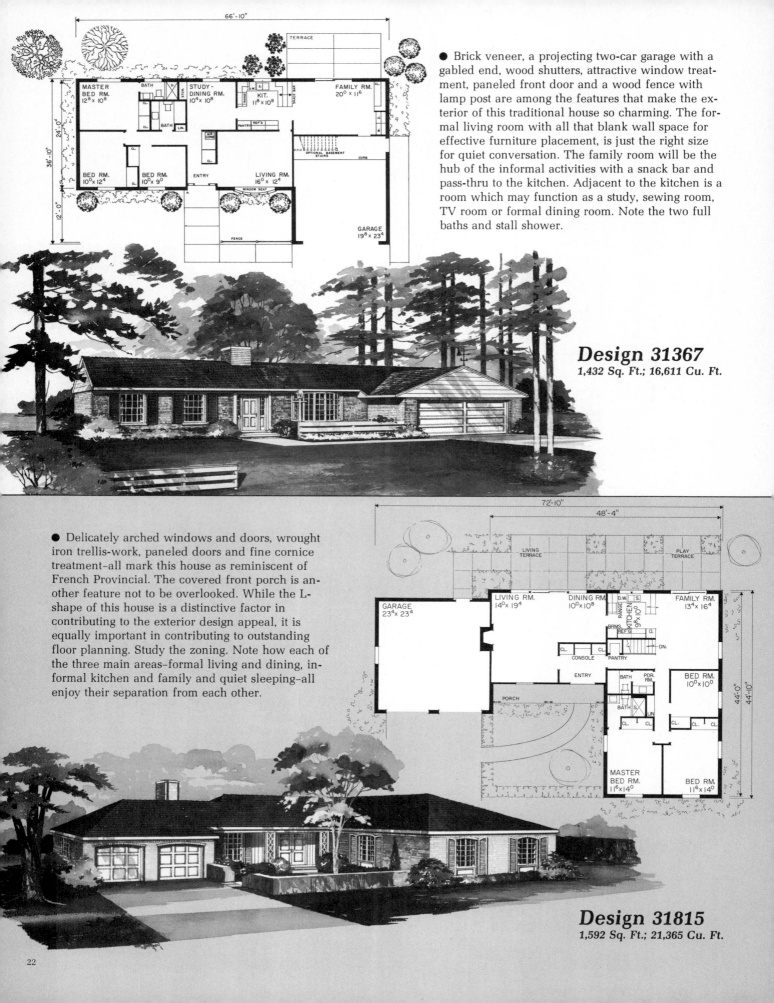

● Brick veneer, a projecting two-car garage with a gabled end, wood shutters, attractive window treatment, paneled front door and a wood fence with lamp post are among the features that make the exterior of this traditional house so charming. The formal living room with all that blank wall space for effective furniture placement, is just the right size for quiet conversation. The family room will be the hub of the informal activities with a snack bar and pass-thru to the kitchen. Adjacent to the kitchen is a room which may function as a study, sewing room, TV room or formal dining room. Note the two full baths and stall shower.

Design 31367
1,432 Sq. Ft.; 16,611 Cu. Ft.

● Delicately arched windows and doors, wrought iron trellis-work, paneled doors and fine cornice treatment–all mark this house as reminiscent of French Provincial. The covered front porch is another feature not to be overlooked. While the L-shape of this house is a distinctive factor in contributing to the exterior design appeal, it is equally important in contributing to outstanding floor planning. Study the zoning. Note how each of the three main areas–formal living and dining, informal kitchen and family and quiet sleeping–all enjoy their separation from each other.

Design 31815
1,592 Sq. Ft.; 21,365 Cu. Ft.

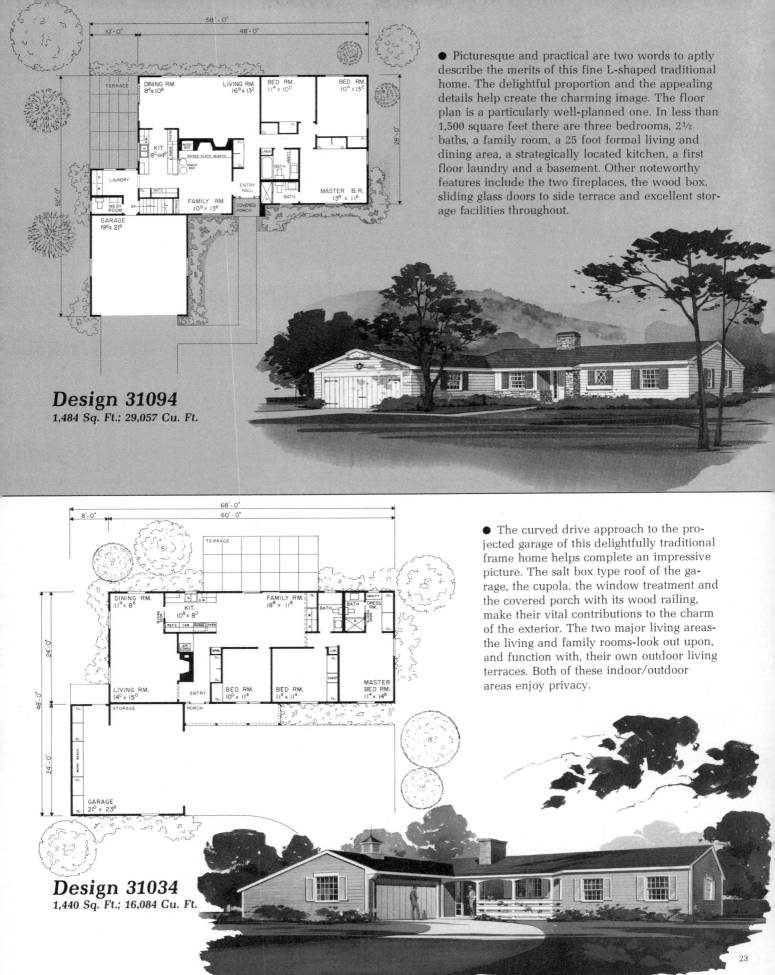

● Picturesque and practical are two words to aptly describe the merits of this fine L-shaped traditional home. The delightful proportion and the appealing details help create the charming image. The floor plan is a particularly well-planned one. In less than 1,500 square feet there are three bedrooms, 2½ baths, a family room, a 25 foot formal living and dining area, a strategically located kitchen, a first floor laundry and a basement. Other noteworthy features include the two fireplaces, the wood box, sliding glass doors to side terrace and excellent storage facilities throughout.

Design 31094
1,484 Sq. Ft.; 29,057 Cu. Ft.

● The curved drive approach to the projected garage of this delightfully traditional frame home helps complete an impressive picture. The salt box type roof of the garage, the cupola, the window treatment and the covered porch with its wood railing, make their vital contributions to the charm of the exterior. The two major living areas-the living and family rooms-look out upon, and function with, their own outdoor living terraces. Both of these indoor/outdoor areas enjoy privacy.

Design 31034
1,440 Sq. Ft.; 16,084 Cu. Ft.

Design 33211
1,344 Sq. Ft.; 26,423 Cu. Ft.

● This is a captivating version of the three bedroom, two-bath L-shaped home. Notice that spacious living-dining area with the many windows to the side plus a view of the rear terrace from the dining room. The living room also has a fireplace for added appeal. The breakfast room also will enjoy the rear terrace and includes such features as a pantry, china cabinet and storage closet. The kitchen is an efficient work center with nearby adjacent basement stairs. The basement could be developed into an informal recreation area.

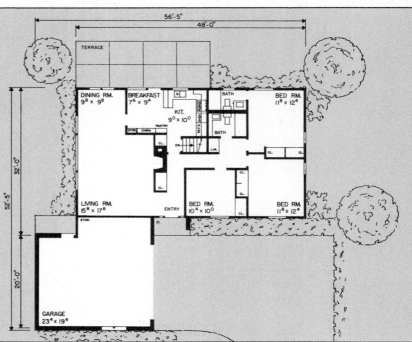

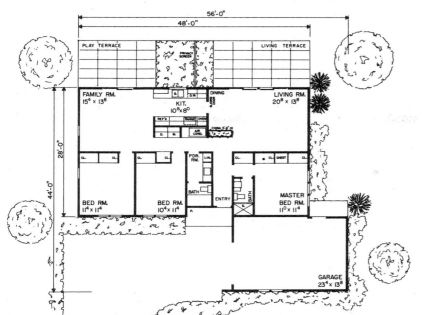

Design 31188
1,326 Sq. Ft.; 16,469 Cu. Ft.

● A charming, traditionally styled, L-shaped home which can be built most economically. Constructed on a concrete slab, the house itself forms a perfect rectangle, thus assuring the most efficient use of materials. The center entrance routes traffic to the various areas of this plan most conveniently. Compact, yet spacious. The three bedrooms are sizable and enjoy their privacy. The entire breadth of the rear experiences the delightful feeling of openness.

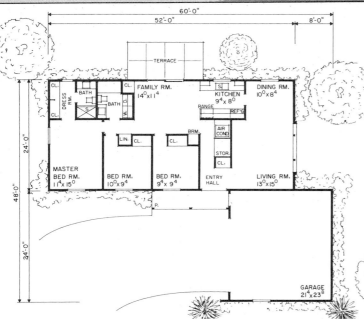

Design 32122 1,248 Sq. Ft.; 14,560 Cu. Ft.

● An ideal home for a family of modest size with a modest budget. This charming frame house can be built most economically. The main portion of the house is a perfect 52 x 24 foot rectangle. The projecting two-car garage with its cupola and traditionally sloped roof, adds immeasurably to the exterior appeal. The double front doors are sheltered by the covered porch. Inside, there is a wealth of livability. Two full baths, located back-to-back for plumbing economy, service the three bedrooms.

OPTIONAL BASEMENT PLAN

Design 31107   1,416 Sq. Ft.; 14,755 Cu. Ft.

● A smart looking traditional adaptation which, because of its perfectly rectangular shape, will be most economical to build. The low-pitched roof has a wide overhang which accentuates its low-slung qualities. The attached two-car garage is oversized to permit the location of extra bulk storage space. Further, its access to the house is through the handy separate laundry area. This house will function as either a four bedroom home, or as one that has three bedrooms, plus a quiet study. Features include a fireplace in the living room, built-in china cabinet in the breakfast room, sizable vanity in the main bath and more.

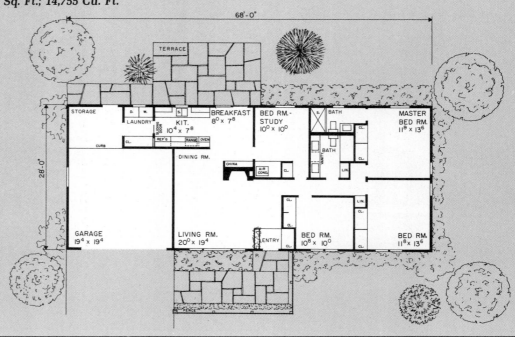

Design 31803
1,434 Sq. Ft.; 28,552 Cu. Ft.

● The texture of brick does much to enhance the beauty of this pleasing traditional home. The center entrance with its double doors is protected by the covered porch. Traffic will flow easily and efficiently. The end-living room will be free of annoying cross-room traffic. A centered fireplace wall with end book shelves is an outstanding feature of this area. The separate dining room is ideally located between the living room and kitchen which allows for its maximum use.

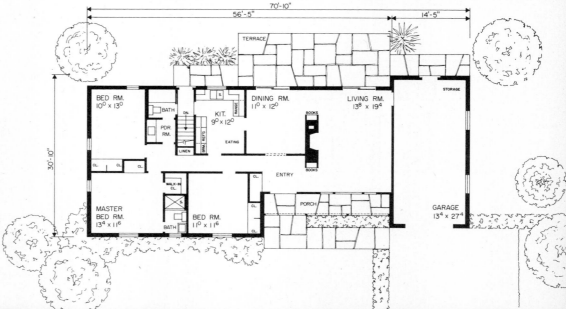

Design 31374   1,248 Sq. Ft.; 25,336 Cu. Ft.

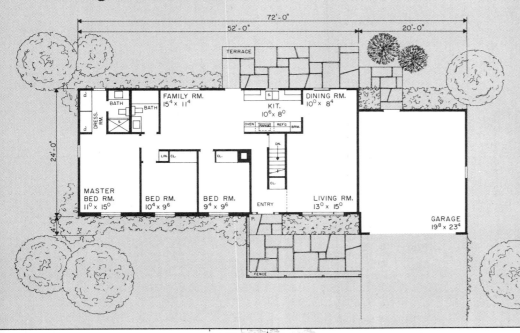

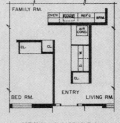

OPTIONAL NO BASEMENT PLAN

● A low-pitched, overhanging roof that hugs the ground and gives an impression that belies the actual dimensions of the home is a feature of this design. Board and batten siding, stone veneer, muntined windows and paneled shutters are all earmarks of Early American styling. Planning is open and pleasant; the living area leads to the dining space and through sliding glass doors to the outdoor terrace; the in-line kitchen leads to the all-purpose family room.

Design 32671
1,589 Sq. Ft.; 36,210 Cu. Ft.

● The rustic exterior of this one-story home features vertical wood siding. The entry foyer is floored with flagstone and leads to the three areas of the plan: sleeping, living and work center. The sleeping area has three bedrooms, the master bedroom has sliding glass doors to the rear terrace. The living area, consisting of gathering and dining rooms, also has access to the terrace. The work center is efficiently planned. It houses the kitchen with snack bar, breakfast room with built-in china cabinet and stairs to the basement. This is a very livable plan.

Design 31088
1,408 Sq. Ft.; 14,305 Cu. Ft.

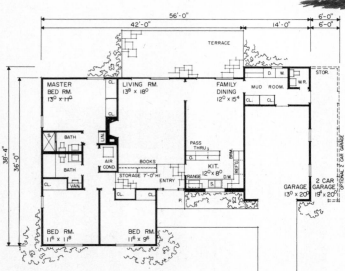

● This pleasing home will be just the one to satisfy the medium budget. You'll forever be proud of what your construction dollar bought; for you'll never tire of the attractive exterior and you'll never cease to be thankful for the fine livability of the interior. The three bedrooms are serviced by two full baths. The family's living habits will be admirably taken care of by the big formal living room and the multi-purpose family room. The homemaker will love the front kitchen and be grateful for the pass-thru to the family room. Don't miss the built-in book storage unit, the bathroom vanity, the extra closets, the wash room and the sliding glass doors.

Traditional and Contemporary
Optional Exteriors & Plans
1344 to 1729 Sq. Ft.

● Here is a unique series of designs with three charming exterior adaptations—Southern Colonial, Western Ranch, French Provincial—and two distinctive floor plans. Each plan has a different design number and is less than 1,600 square feet.

If yours is a preference for the floor plan featuring the 26 foot keeping room, you should order blueprints for Design 32611. Of course, the details for each of the three delightful exteriors will be included. On the other hand, should the plan with the living, dining and family rooms be your favorite, order blueprints for Design 32612 and get details for all three exteriors.

There are many points of similarity in the two designs. Each has a fireplace, 2½ baths, sliding glass doors to the rear terrace, master bedroom with walk-in closet and private bath with stall shower and a basement. It is interesting to note that two of the exteriors have covered porches. Don't miss the beamed ceilings, the various storage facilities and the stall showers.

Design 32611
1,557 Sq. Ft.; 26,245 Cu. Ft.

Design 32612
1,571 Sq. Ft.; 32,880 Cu. Ft.

Design 31305 1,382 Sq. Ft.; 16,584 Cu. Ft.

● Order blueprints for any one of the three exteriors shown on this page and you will receive details for building this outstanding floor plan. You'll find the appeal of these exteriors difficult to beat. As for the plan, in less than 1,400 square feet there are three bedrooms, two full baths, a separate dining room, a formal living room, a fine kitchen overlooking the rear yard and an informal family room. In addition, there is the attached two-car garage. Note the location of the stairs when this plan is built with a basement. Each of the exteriors is predominantly brick - the front of Design 31305 (above) features both stone and vertical boards and battens with brick on the other three sides. Observe the double front doors of the French design, 31382 (below) and the Contemporary design, 31383 (bottom). Study the window treatment.

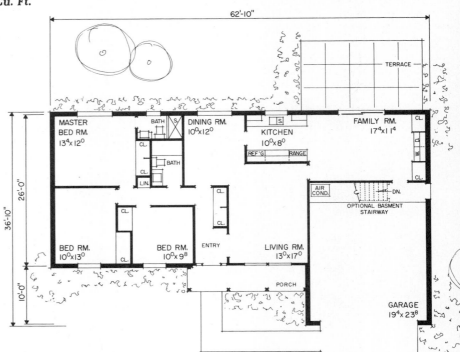

Design 31382
1,382 Sq. Ft.; 17,164 Cu. Ft.

Design 31383
1,382 Sq. Ft.; 15,448 Cu. Ft.

Design 31307 1,357 Sq. Ft.; 14,476 Cu. Ft.

● These three stylish exteriors have the same practical, L-shaped floor plan. Design 31307 (above) features a low-pitched, wide-overhanging roof, a pleasing use of horizontal siding and brick and an enclosed front flower court. Design 31380 (below) has its charm characterized by the pediment gables, the effective window treatment and the masses of brick. Design 31381 (bottom) is captivating because of its hip-roof, its dentils, panelled shutters and lamp post. Each of these three designs has a covered front porch. Inside, there is an abundance of livability. The formal living and dining area is spacious, and the U-shaped kitchen is efficient. There is informal eating space, a separate laundry and a fine family room. Note the sliding glass doors to the terrace. The blueprints include details for building either with or without a basement. Observe the pantry of the non-basement plan.

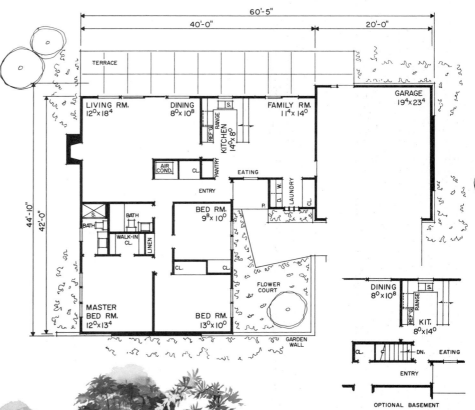

Design 31380
1,399 Sq. Ft.; 17,709 Cu. Ft.

Design 31381
1,399 Sq. Ft.; 17,937 Cu. Ft.

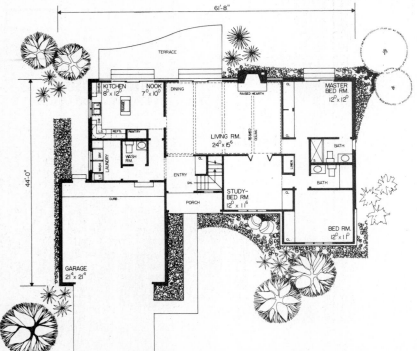

Design 32565
1,540 Sq. Ft.; 33,300 Cu. Ft.

● This modest sized floor plan has much to offer in the way of livability. It may function as either a two or three bedroom home. The living room is huge and features a fine, raised hearth fireplace. The open stairway to the basement is handy and will lead to what may be developed as the recreation area. In addition to the two full baths, there is an extra wash room. Adjacent is the laundry room and the service entrance from the garage. The blueprints you order for this design will show details for each of the three delightful elevations above. Which is your favorite? The Tudor, the Colonial or the Contemporary?

Design 31323 1,344 Sq. Ft.; 17,472 Cu. Ft.

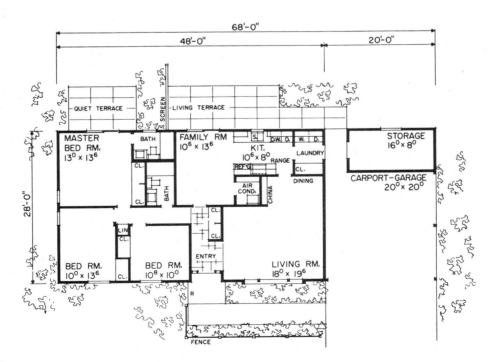

● Incorporated in each set of blueprints for this design are details for building each of the three charming, traditional exteriors. Each of the three alternate exteriors has a distinction all its own. A study of the floor plan reveals fine livability. There are two full baths, a fine family room, an efficient work center, a formal dining area, bulk storage facilities and sliding glass doors to the quiet and living terraces. Laundry is strategically located near the kitchen.

Design 31389
1,488 Sq. Ft.; 18,600 Cu. Ft.

● Your choice of exterior goes with the outstanding floor plan on the opposing page. If your tastes include a liking for French Provincial, Design 31389, above, will provide a lifetime of satisfaction. On the other hand, should you prefer the simple straightforward lines of comtemporary design, the exterior for Design 31387, below, will be your favorite. For those who enjoy the warmth of Colonial adaptations, the charming exterior for Design 31388, on the opposite page, will be perfect. Of interest, is a comparison of these three exteriors. Observe the varying design treatment of the windows, the double front doors, the garage doors and the roof lines. Don't miss other architectural details. Study each exterior and the floor plan carefully. Three charming designs you won't want to miss.

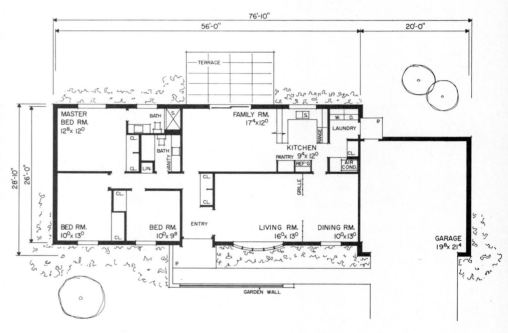

Design 31387
1,488 Sq. Ft.; 16,175 Cu. Ft.

Design 31388
1,488 Sq. Ft.; 18,600 Cu. Ft.

Design 31864
1,598 Sq. Ft.; 27,611 Cu. Ft.

● What's your favorite exterior? The one above which has a distinctive colonial appearance, or that below with its sleek contemporary look? Maybe you prefer the more formal hip-roof exterior (bottom) with its French feeling. Whatever your choice, you'll know your next home will be one that is delightfully proportioned and is sure to be among the most attractive in the neighborhood. It is interesting to note that each exterior highlights an effective use of wood siding and stone (or brick, as in the case of Design 31866). The floor plan features three bedrooms, 2½ baths, a formal living and dining room, a snack bar and a mud room. The master bedroom of the contemporary design has its window located in the left side elevation wall.

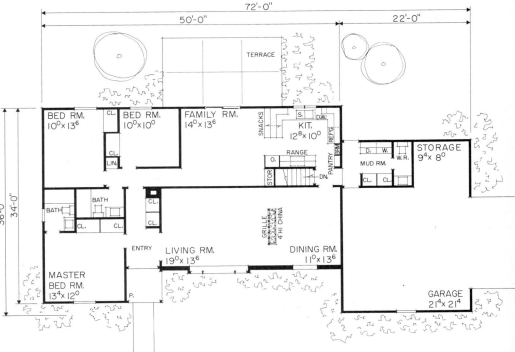

Design 31865
1,598 Sq. Ft.; 25,626 Cu. Ft.

Design 31866
1,598 Sq. Ft.; 27,248 Cu. Ft.

Design 32810
3-Bedroom Plan

Design 32814
4-Bedroom Plan

1,536 Sq. Ft.; 34,560 Cu. Ft.

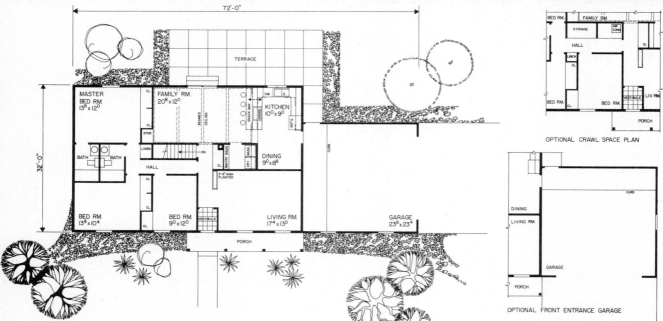

- 2 x 6 stud wall construction front and center! The designs on these two pages are particularly energy-efficient minded. All exterior walls employ the use of the larger size stud (in preference to the traditional 2 x 4 stud) to permit the installation of extra thick insulation. The high cornice design also allows for more ceiling insulation. In addition to the insulation factor, 2 x 6 studs are practical from an economic standpoint. According to many experts, the use of 2 x 6's spaced 24 inches O.C. results in the need for less lumber and saves construction time. However, the energy-efficient features of this series do not end with the basic framing members. Efficiency begins right at the front door where the vestibule acts as an airlock restricting the flow of cold air to the interior. The basic rectangular shape of the house spells efficiency. No complicated and costly construction here. Yet, there has been no sacrifice of delightful exterior appeal. Efficiency and economy are also embodied in such features as back-to-back plumbing, centrally located furnace, minimal window and door openings and, most important of all - size.

Design 32811
3-Bedroom Plan

Design 32815
4-Bedroom Plan

1,581 Sq. Ft.; 36,694 Cu. Ft.

Design 32812
3-Bedroom Plan
Design 32816
4-Bedroom Plan

1,581 Sq. Ft.; 35,040 Cu. Ft.

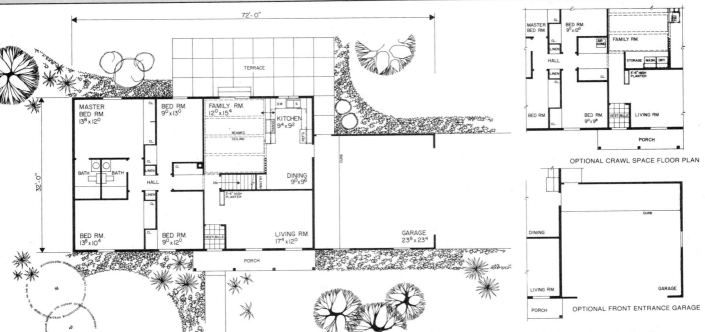

Within 1,536 square feet there is outstanding livability and a huge variety of options from which to choose. For instance, of the four stylish exteriors, which is your favorite? The cozy, front porch Farmhouse adaptation; the pleasing Southern Colonial version, the French creation, or the rugged Western facade? Further, do you prefer a three or a four bedroom floor plan? With or without a basement? Front or side-opening garage? If you wish to order blueprints for the hip-roofed design with three bedrooms, specify Design 32812; for the four bedroom option specify 32816. To order blueprints for the three bedroom Southern Colonial, request Design 32811; for the four bedroom model, ask for Design 32815, etc. All blueprints include the optional non-basement and front opening garage details. Whatever the version you select, you and your family will enjoy the beamed ceiling of the family room, the efficient, U-shaped kitchen, the dining area, the traffic-free living room and the fine storage facilities. Truly, a fine design series created to give each home buyer the maximum amount of choice and flexibility.

Design 32813
3-Bedroom Plan
Design 32817
4-Bedroom Plan

1,536 Sq. Ft.; 33,334 Cu. Ft.

Design 32825
1,584 Sq. Ft.; 37,215 Cu. Ft.

● With today's tight economy, this house will be a real bargain. It has all of the necessary features to insure gracious living yet keep costs down - generous living space, packed with amenities and constructed with durable materials. Locating the garage to the front of this design is practical because it makes the overall width only 51' so it will fit on a narrow lot and it will act as a buffer against street noise. The interior of this home will be interesting. Three sets of sliding glass doors at the rear of the plan will flood the interior with natural light. Since it is a modified open plan it will allow the sunlight to penetrate deep into the interior. The gathering room which seems expanded by the cathedral ceiling has a fireplace.

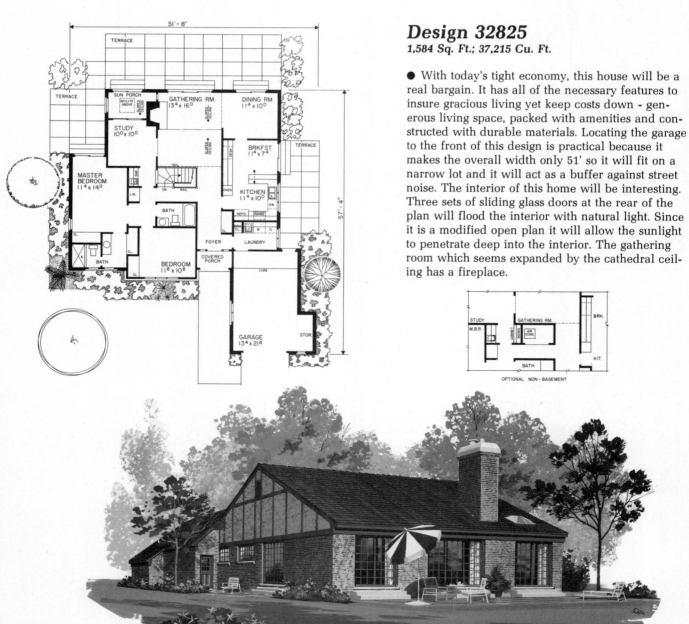

Contemporary
One-Story Homes
1200 - 1599 Sq. Ft.

Design 32818 1,566 Sq. Ft.; 20,030 Cu. Ft.

● This is most certainly an outstanding contemporary design. Study the exterior carefully before your journey to inspect the floor plan. The vertical lines are carried from the siding to the paned windows to the garage door. A overhanging hip-roof. The front entry is recessed so the overhanging roof creates a covered porch. Note the planter court with privacy wall. The floor plan is just as outstanding. The rear gathering room has a sloped ceiling, raised hearth fireplace, sliding glass doors to the terrace and a snack bar with pass-thru to the kitchen. In addition to the gathering room, there is the living room/study. This room could be utilized in a variety of ways. The formal dining room is convenient to the U-shaped kitchen. Three bedrooms and two closely located baths are in the sleeping wing. This plan includes details for the construction of an optional basement.

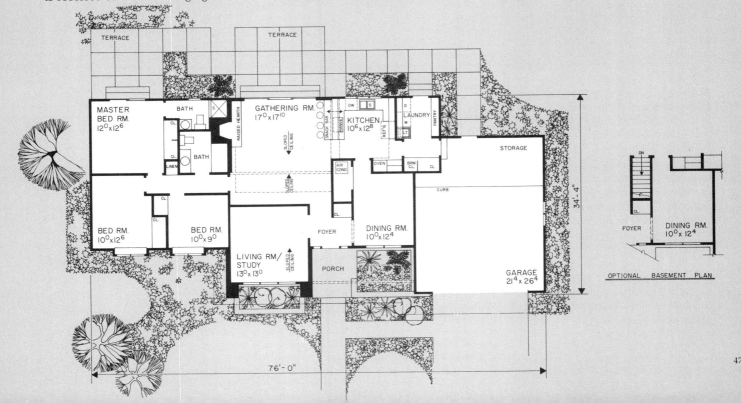

47

- A well-planned, medium sized contemporary home with plenty of big-house features. The brick line of the projected bedroom wing extends toward the projected two-car garage to form an attractive front court. The large glass panels below the overhanging roof are a dramatic feature. In addition to the two full baths, there is an extra wash room easily accessible from the kitchen/family room area as well as the outdoors. The laundry equipment is strategically located.

Design 31342
1,560 Sq. Ft.; 13,256 Cu. Ft.

- Here is a relatively low-cost home with a majority of the features found in today's high priced homes. The three-bedroom sleeping area highlights two full baths. The living area is a huge room of approximately 25 feet in depth zoned for both formal living and dining activities. The kitchen is extremely well-planned and includes a built-in desk and pantry. The family room has a snack bar and sliding glass doors to the terrace. Blueprints include optional basement details.

Design 31357
1,258 Sq. Ft.; 13,606 Cu. Ft.

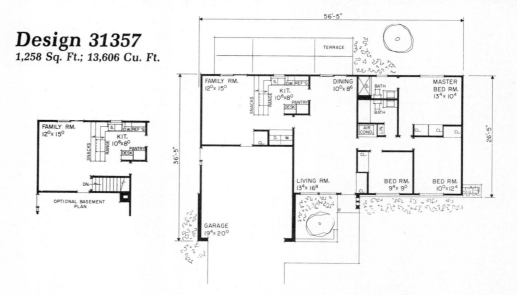

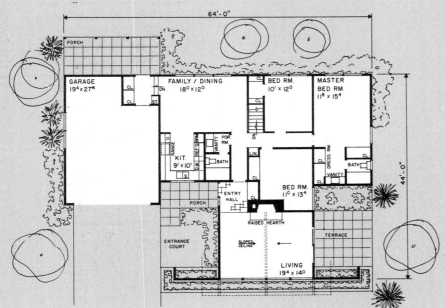

Design 31065
1,492 Sq. Ft.; 24,407 Cu. Ft.

● Here is a refreshing design that reflects all that is so appealing in good up-to-date exterior detailing and practical, efficient floor planning. The low pitched, wide overhanging roof, the glass gabled end with exposed rafters, the raised planter and the extended wing walls are all delightful exterior features. Of particular interest, is the formation of the front entrance court and side terrace resulting from the extension of the front living room wall. Inside, there is much to excite the new occupants. The quietly, formal living room will have plenty of light. The strategically located kitchen is ideal to view approaching callers. Master bedroom with bath, vanity and dressing room housing two closets will be welcomed by the parents.

49

Design 33190
1,516 Sq. Ft.; 29,280 Cu. Ft.

● Imagine the pleasing contemporary living patterns to be enjoyed in this unique, eye-catching house. Here are 1,516 well-planned square feet. The living and dining areas are cheerful and spacious. Large glass panels permit a view to the front of the site, while sliding glass doors open onto the covered play porch. The massive chimney services the raised hearth fireplace of the living room and the barbecue unit of the porch. The fine kitchen has eating space and functions in a most convenient fashion with the pool terrace. The three bedrooms are served by two full baths. The main bath has a compartmented powder room. The master bath's stall shower is accessible from the pool.

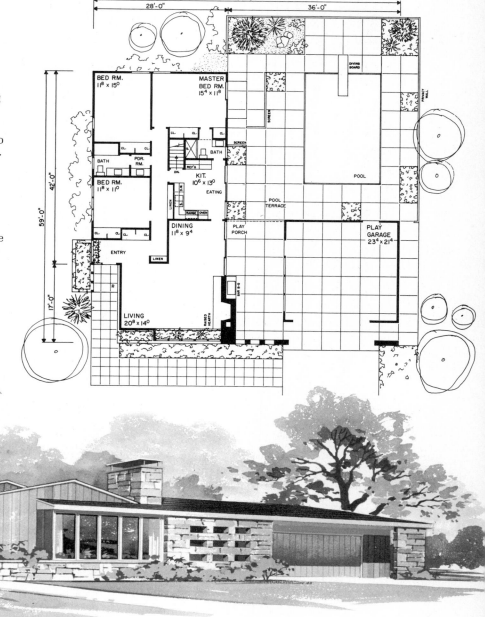

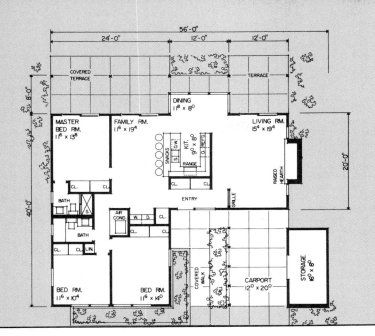

Design 31319 1,492 Sq. Ft.; 15,183 Cu. Ft.

● Here's a cleverly-designed contemporary home that gives you an absolute minimum of wasted space inside, and even puts the open portion of the L-shape to good use. You might expect such a tightly-designed house to sacrifice good traffic and zoning, but this home doesn't. The children's two front bedrooms are throughly sound-conditioned from noise in the living areas, and there's a chance for peace in the living room when the youngsters watch TV in the family room. An interior U-shaped kitchen frees-up valuable outside wall space. The snack bar will be great for those informal meals. Indoor/outdoor enjoyment can be achieved in all the major living areas - the family room, dining room and living room - and the master bedroom.

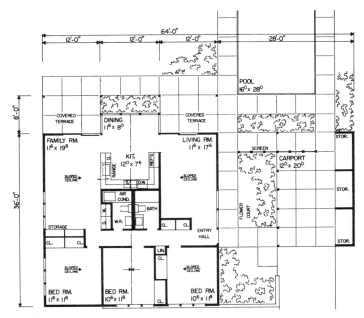

Design 31182
1,350 Sq. Ft.; 13,716 Cu. Ft.

● If you are contemporary-minded and your building budget is minimal you need not look any further or wait any longer. This basically rectangular house has all the interest of one twice its size, for the subtleties of good design are many. Floor planning a small home is often very difficult. Yet, rarely has a small home been better planned. Each of the rooms enjoys privacy from the other and traffic patterns are flexible. Observe the economies resulting from the two full bath's back-to-back plumbing located in the center of the plan. The three front bedrooms have sloped ceilings to highlight their appeal. The family room and living room also have the attraction of sloped ceilings.

Design 32821
1,363 Sq. Ft. – First Floor
357 Sq. Ft. – Second Floor
37,145 Cu. Ft.

Mansard Roof Adaptation

A Trend House . . .

● Here is a truly unique house whose interior was designed with the current decade's economies, lifestyles and demographics in mind. While functioning as a one-story home, the second floor provides an extra measure of livability when required. In addition, this two-story section adds to the dramatic appeal of both the exterior and the interior. Within only 1,363 square feet, this contemporary delivers refreshing and outstanding living patterns for those who are buying their first home, those who have raised their family and are looking for a smaller home and those in search of a retirement home. The center entrance routes traffic effectively to each area. The great room with its raised hearth fireplace, two-story arching and delightful glass areas is most distinctive. The kitchen is efficient and but a step from the dining room. The covered porch will provide an ideal spot for warm-weather, outdoor dining. The separate laundry room is strategically located. The sleeping area may consist of one bedroom and a study, or two bedrooms. Each room functions with the sheltered wood deck - a perfect location for a hot tub.

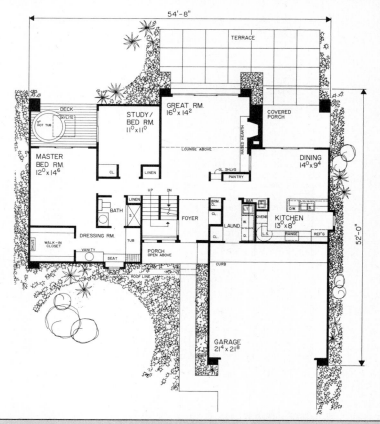

Design 32822
1,363 Sq. Ft. – First Floor
351 Sq. Ft. – Second Floor
36,704 Cu. Ft.

Gable Roof Version

● The full bath is planned to have easy access to the master bedroom and living areas. Note the stall shower, tub, seat and vanity. The second floor offers two optional layouts. It may serve as a lounge, studio or hobby area overlooking the great room. Or, it may be built to function as a complete private guest room. It would be a great place for the visiting grandchildren. Don't miss the outdoor balcony. Additional livability and storage facilities may be developed in the basement. Then, of course, there are two exteriors to choose from. Design 32821, with its horizontal frame siding and deep, attractive cornice detail, is an eye-catcher. For those with a preference for a contemporary fashioned gable roof and vertical siding, there is Design 32822. With the living areas facing the south, these designs will enjoy benefits of passive solar exposure. The overhanging roofs will help provide relief from the high summer sun. This is surely a modest-sized floor plan which will deliver new dimensions in small-family livability.

... For the 80's and Decades to Come

Design 33203

1,291 Sq. Ft.; 23,625 Cu. Ft.

● The degree to which your family's living patterns will be efficient will depend upon the soundness of the basic floor plan. Here is an exceptionally practical arrangement for a medium sized home. Traffic patterns will be most flexible. The work area is strategically located close to the front door and it will function ideally with both the indoor and outdoor informal living areas. The master bath will serve the living room and the sleeping area. The second bath will serve the work area, the family room and third bedroom. Note stall shower.

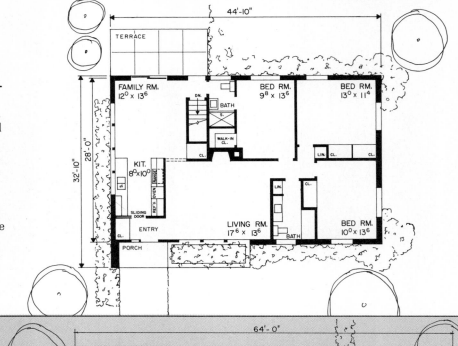

Design 33212

1,493 Sq. Ft.; 16,074 Cu. Ft.

● Imagine ushering your first visitors through the front door of your new home. After you have placed their coats in the big closet in the front entry, you will show them your three sizable bedrooms and the two full baths. Next, you will take them through the separate dining area, the efficient kitchen and into the large family room with its wall of built-in storage units. Your visitors will comment about the sliding glass doors which open onto the terrace from the family room.

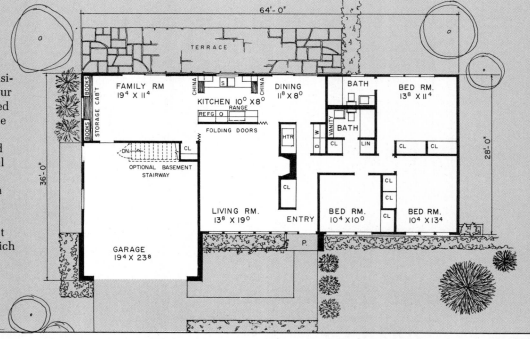

Design 31072

1,232 Sq. Ft.; 22,484 Cu. Ft.

● Low-pitched overhanging roof, vertical siding and patterned masonry screen wall create a charming exterior for this house. A fireplace divides the living room from the entry hall. Seven-foot high cabinets separate the kitchen from the living/dining area. The dining area opens through sliding glass doors to a private covered porch and the rear terrace. The informal area is enhanced by a wall of windows overlooking the terrace and a three-foot high built-in planter. For recreation and bulk storage there is a full basement.

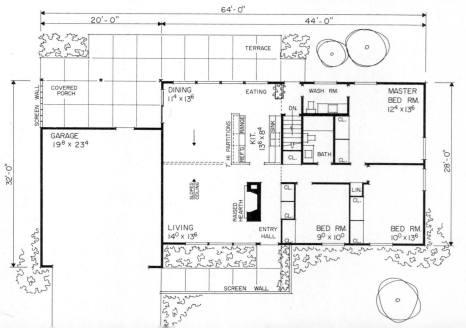

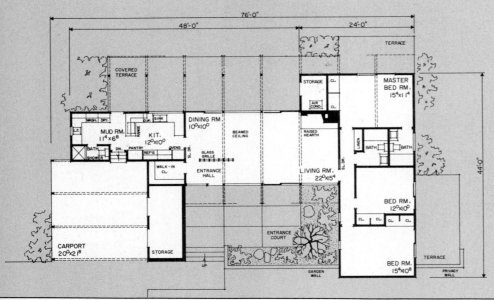

● Lovers of good contemporary design backed-up by completely modern and practical floor planning had better take note. This flat-roofed design is a good study in masses. There is the mass of vertical wood siding, the mass of brick veneer and the mass of glass panels. All these areas add to the dramatic simplicity of the facade. The wide steps leading to the raised entrance court are an inviting feature. The exposed beams, visible from the front court, run right through the house and out onto the rear terrace. Also a covered terrace.

Design 31276
1,286 Sq. Ft.; 12,860 Cu. Ft.

● Despite the tropical setting, here is an attractive contemporary design which serves its occupants ideally, whatever the location. Designed for the retired couple or the small family, this home's H-shaped floor plan will result in convenient and interesting living patterns. The kitchen area overlooks the rear covered terrace and separates the sleeping and living wings. Each of the two bedrooms features a wonderful storage wall which includes two closets and a built-in clothes chest. The front living room will be quiet indeed, for it will be free of annoying traffic patterns. The family/dining area has a raised hearth fireplace. Note work shop in the garage with wash room.

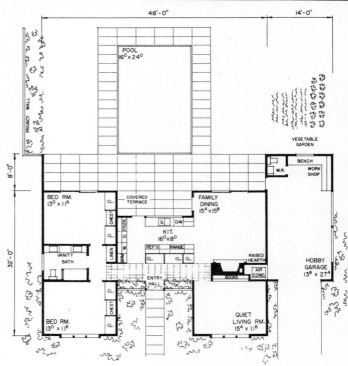

Design 31922
1,573 Sq. Ft.; 14,472 Cu. Ft.

Design 31314
1,600 Sq. Ft.; 16,075 Cu. Ft.

● Low-pitched, overhanging roofs and vertical boards and battens highlight the interior of this modified U-shaped ranch home. A raised stone planter helps create a delightful front entrance. Large glass areas permit the enjoyment of the outdoors from the dining and living rooms. A two-way fireplace services both of these areas simultaneously. The sleeping wing features three bedrooms, two full baths and fine closet facilities. The work center is most efficient. The kitchen is flanked by the laundry with its adjacent wash room and the family room. The outdoor terrace is accessible from both the family and living rooms through sliding glass doors. Note optional basement.

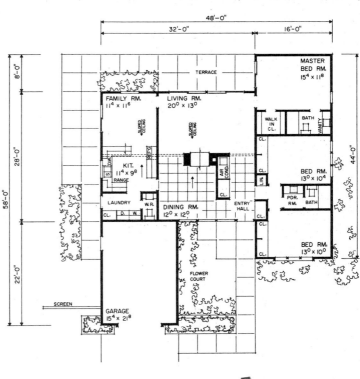

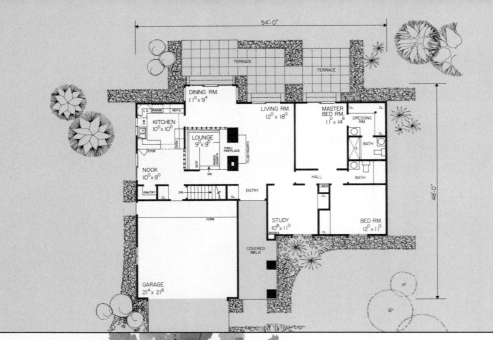

Design 32703
1,445 Sq. Ft.; 30,300 Cu. Ft.

● This modified, hip-roofed contemporary design will be the answer for those who want something both practical, yet different, inside and out. The covered front walk sets the stage for entering a modest sized home with tremendous livability. The focal point will be the pleasant conversation lounge. It is sunken, partically open to the other living areas and shares the enjoyment of the thru-fireplace with the living room. There are two bedrooms, two full baths and a study. The kitchen is outstanding.

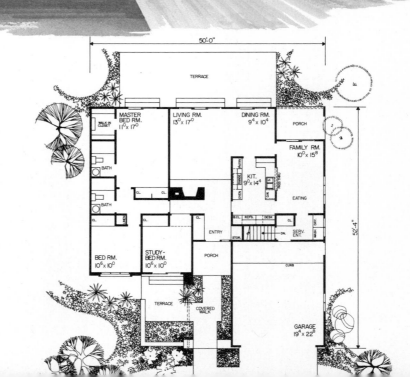

Design 32753
1,539 Sq. Ft.; 31,910 Cu. Ft.

● In this day and age of expensive building sites, projecting the attached garage from the front line of the house makes a lot of economic sense. It also lends itself to interesting roof lines and plan configurations. Here, a pleasing covered walkway to the front door results. A privacy wall adds an extra measure of design appeal and provides a sheltered terrace for the study/bedroom. You'll seldom find more livability in 1,539 square feet. Imagine, three bedrooms, two baths, a spacious living/dining area and a family room.

Design 32744
1,381 Sq. Ft.; 17,530 Cu. Ft.

● Here is a practical and an attractive contemporary home for that narrow building site. It is designed for efficiency with the small family or retired couple in mind. Sloping ceilings foster an extra measure of spaciousness. In addition to the master bedroom, there is the study that can also serve as the second bedroom or as an occasional guest room. The single bath is compartmented and its dual access allows it to serve living and sleeping areas more than adequately. Note raised hearth fireplace, snack bar, U-shaped kitchen, laundry, two terraces, etc.

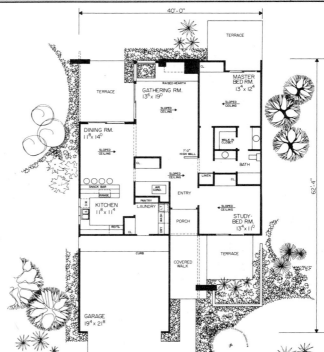

Design 31195

1,344 Sq. Ft.; 12,983 Cu. Ft.

● This is not a large home, yet the convenience and livability it offers is great indeed. It is basically a perfect rectangle with the carport projecting to the front to achieve a delightful feeling of distinctiveness. The carport's flat roof extends to provide an interesting covered walk to the front door. The three bedrooms are located to the front of the house thus allowing for the full enjoyment of the rear terraces. Two baths, situated back-to-back, are handy to the family room and the outdoor terrace. Details for building this design with a basement are also included in the blueprints.

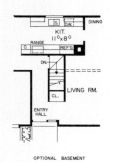

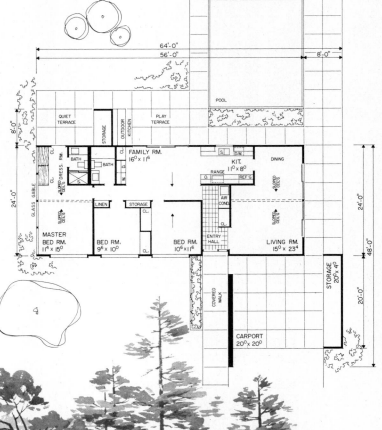

Design 31073

1,386 Sq. Ft.; 13,959 Cu. Ft.

● The refreshing appeal of this contemporary home can be traced to its low-pitched, wide overhanging roofs and the flat roof of the carport which extends over the front privacy wall to help form an enclosed entrance court. Such a delightful area will permit wonderful outdoor living right in the front yard while being protected from the devious eyes of passers-by. This enclosed court also will be enjoyed from within. The large glass area of the living room will permit a view of the dramatic outdoor planting areas. The brick fireplace wall and the six-foot planter/china storage unit further enhances the appeal of the living room area. The family room functions ideally with the kitchen and the rear terrace. Plenty of storage areas throughout.

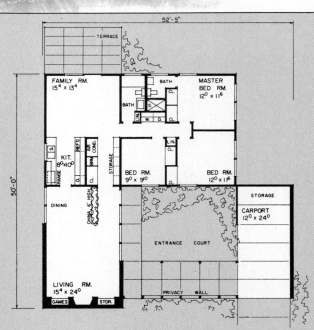

Design 32310
1,435 Sq. Ft.; 14,350 Cu. Ft.

● Here is an attractive low-slung contemporary design whose basic livability is within a 25 x 56 foot rectangle. There are 1,435 square feet and not a single foot of wasted space. The addition of the large porch between house and carport enhances the informal living potential. (Don't miss the built-in barbecue unit.) You may wish to make this area into a family room. The carport completes the L-shaped configuration and provides two large bulk storage areas. The living/dining area with its raised hearth fireplace, sloped ceiling and glass areas, is spacious indeed.

61

Design 33207
1,460 Sq. Ft.; 25,068 Cu. Ft.

● Four bedrooms to satisfy the sleeping requirements of the large or growing family. In addition, there are two full baths nearby. They have a common plumbing wall to enable the installation of these two baths to represent a worthwhile savings. The bath for the master bedroom is also handy to the other three bedrooms. It has a built-in vanity. The main bath is strategically located only a few steps from the kitchen as well as the living area. The kitchen is conveniently located near the front door with the kitchen window overseeing the front yard. The living/dining room will enjoy the fireplace wall. Along with the living/dining room, the family room has sloped ceiling and access to the covered terrace.

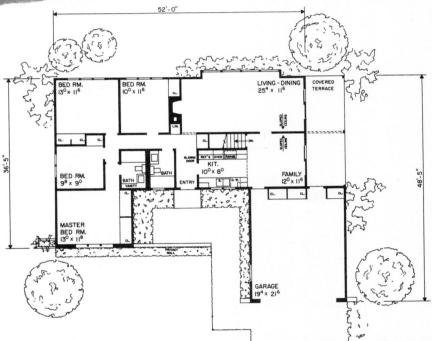

Design 31943 1,565 Sq. Ft.; 17,985 Cu. Ft.

● This home will be a great investment. It starts right out representing economy because it will not require a wide expensive building site. The projecting two-car garage in no way affects the modest, overall, 50 foot width of the house. The favorable economics go even further. In less than 1,600 square feet there is a long list of convenient living features. They begin, of course, with the four sizable bedrooms. Two full baths with back-to-back plumbing will adequately serve the large family. A spacious living room overlooks the rear terrace and has a glass grille room divider. It also features fine wall space for flexible furniture placement. The informal, all-purpose family room functions with the kitchen and terrace.

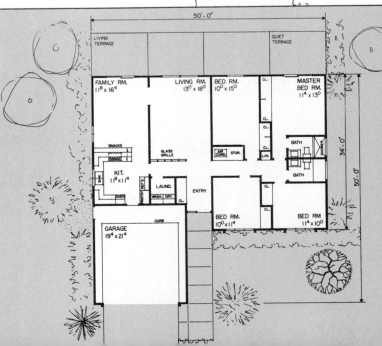

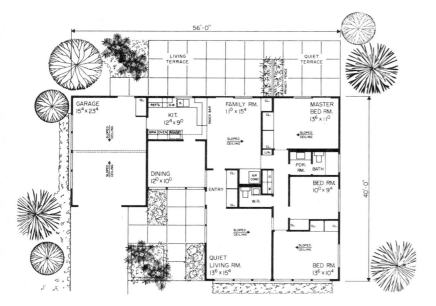

Design 31057
1,320 Sq. Ft.; 13,741 Cu. Ft.

● Here is a relatively small contemporary with 1,320 square feet that is just loaded with livability. Its overall dimension of 56 feet means that it won't require a big, expensive piece of property either. The flexibility of the floor planning is great, too. Notice the room locations. The three bedrooms and compartmented bath occupy the far end of the house. The living room is by itself and will have privacy. The dining room is next to the kitchen and enjoys a view of the front court area. The family room functions ideally with the kitchen and the rear terrace. Other features include the wash room, the sloped ceilings, the snack bar, the attached garage, sliding glass doors and an abundance of closets.

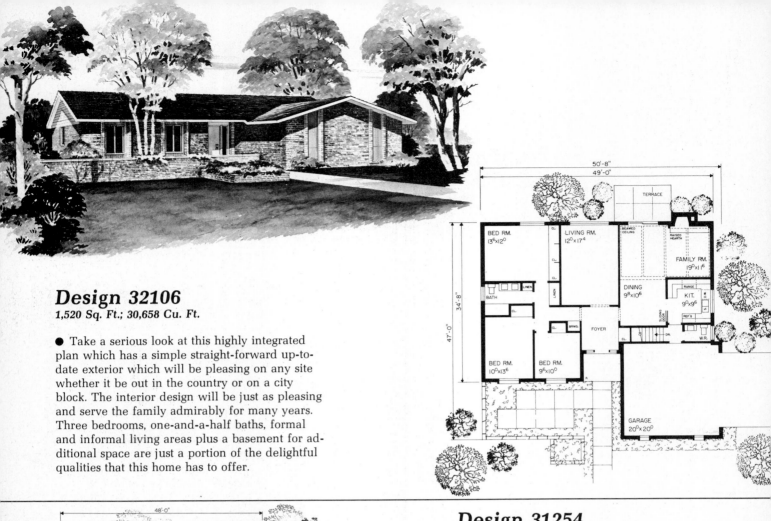

Design 32106
1,520 Sq. Ft.; 30,658 Cu. Ft.

● Take a serious look at this highly integrated plan which has a simple straight-forward up-to-date exterior which will be pleasing on any site whether it be out in the country or on a city block. The interior design will be just as pleasing and serve the family admirably for many years. Three bedrooms, one-and-a-half baths, formal and informal living areas plus a basement for additional space are just a portion of the delightful qualities that this home has to offer.

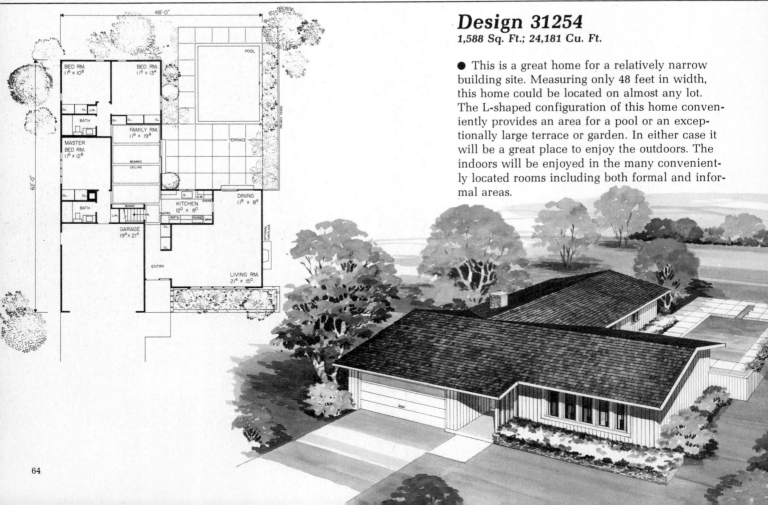

Design 31254
1,588 Sq. Ft.; 24,181 Cu. Ft.

● This is a great home for a relatively narrow building site. Measuring only 48 feet in width, this home could be located on almost any lot. The L-shaped configuration of this home conveniently provides an area for a pool or an exceptionally large terrace or garden. In either case it will be a great place to enjoy the outdoors. The indoors will be enjoyed in the many conveniently located rooms including both formal and informal areas.

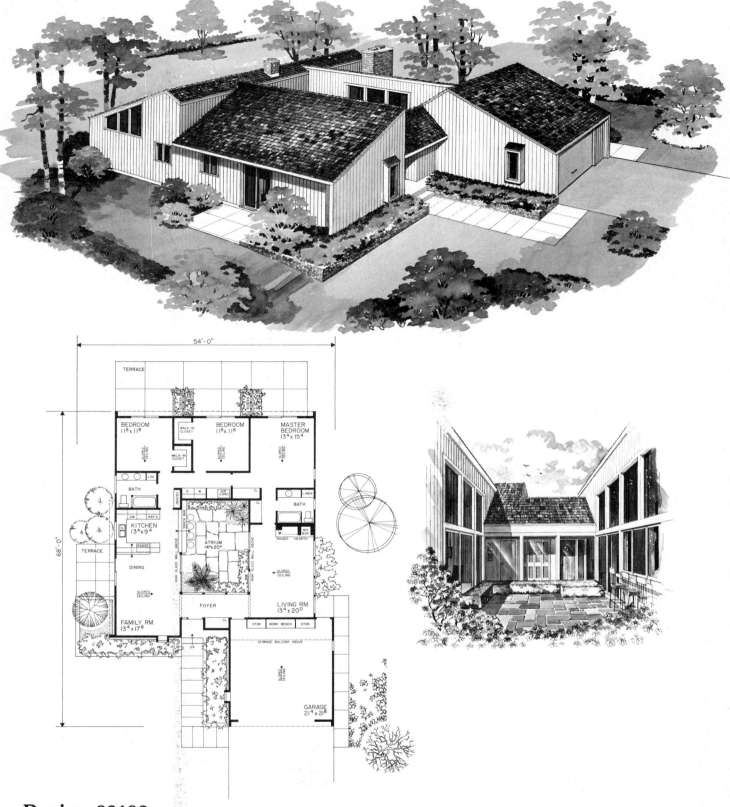

Design 32182 1,558 Sq. Ft.; 280 Sq. Ft. - Atrium; 18,606 Cu. Ft.

● What a great new dimension in living is represented by this unique contemporary design! Each of the major zones comprise a separate unit which, along with the garage, clusters around the atrium. High sloped ceilings and plenty of glass areas assure a feeling of spaciousness. The quiet living room will enjoy its privacy, while activities in the informal family room will be great fun functioning with the kitchen. A snack bar opens the kitchen to the atrium. The view, above right, shows portions of snack bar and the front entry looking through the glass wall. There are two full baths strategically located to service all areas conveniently. Storage facilities are excellent, indeed. Don't miss the storage potential found in the garage. There is a work bench and storage balcony above.

Design 31033
1,500 Sq. Ft.; 15,375 Cu. Ft.

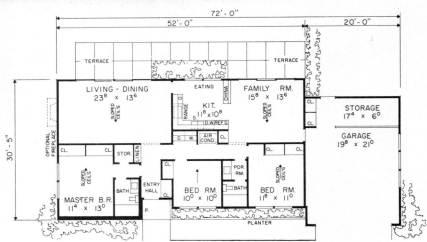

● A unique plan with each of the three bedrooms, plus two full baths, located in the front. This results in the living areas and the kitchen overlooking and functioning with the rear terrace. The master bedroom has its own bath and fine closet facilities. The living room has excellent furniture placement possibilities. A fine feature. Optional basement details are included with the plan.

Design 33163
1,552 Sq. Ft.; 16,171 Cu. Ft.

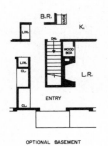

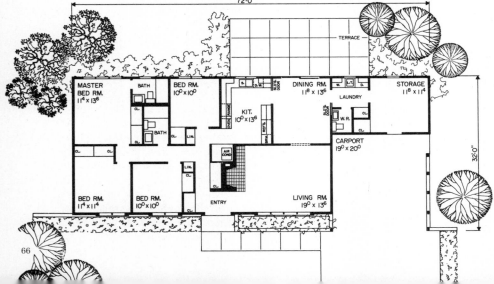

● Four bedrooms and two full baths will very adequately serve the growing family occupying this appealing contemporary. Its perfectly rectangular shape means economical construction. Note the attractive built-in planter adjacent to the front door. The large storage area behind the carport will solve any storage problems. Laundry and wash room are strategically located to serve the family.

Design 31021
1,432 Sq. Ft.; 14,549 Cu. Ft.

OPTIONAL BASEMENT PLAN

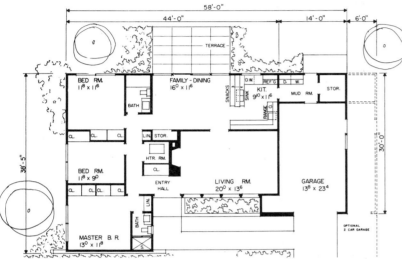

● Behind the double front doors of this straight-forward, contemporary design there is a heap of living to be enjoyed. The large living room with its dramatic glass wall and attractive fireplace will never fail to elicit comments of delight. The master bedroom has a whole wall of wardrobe closets and a private bath. Another two bedrooms and a bath easily serve the family.

Design 31000
1,488 Sq. Ft.; 13,392 Cu. Ft.

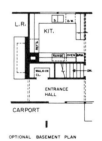

OPTIONAL BASEMENT PLAN

● Living in this flat-roofed contemporary will be lots of fun. And little wonder when you consider the two large living areas, the outstanding kitchen, the zoning of the sleeping area, the outdoor terraces and the spacious front entrance court. Also included in this plan are two full baths, an attractive fireplace and a big carport.

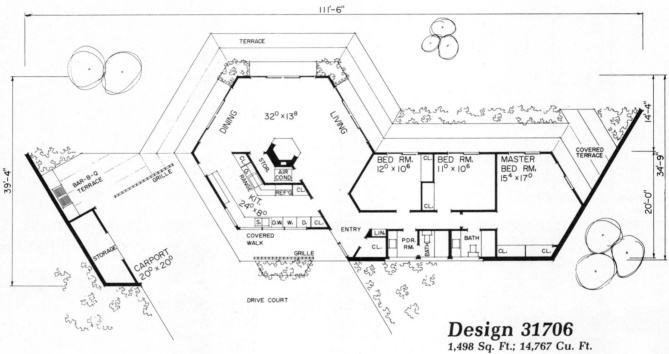

Design 31706
1,498 Sq. Ft.; 14,767 Cu. Ft.

● Contemporary living can adapt itself to many unique and delightful shapes. Whether chosen as a year 'round home in suburbia, or as a leisure-time home at the lake, seashore or in the woods, this home will be delightfully different and fun to live in. The living portion is a hexagon with its hexagonal fireplace the focal point. Observe the economical use of space and efficiency of the kitchen. The projecting sleeping wing contains three bedrooms and two baths. Glass sliding doors open from each room onto a terrace area partially protected by the wide overhanging roof. The carport balances the house and has a side storage unit.

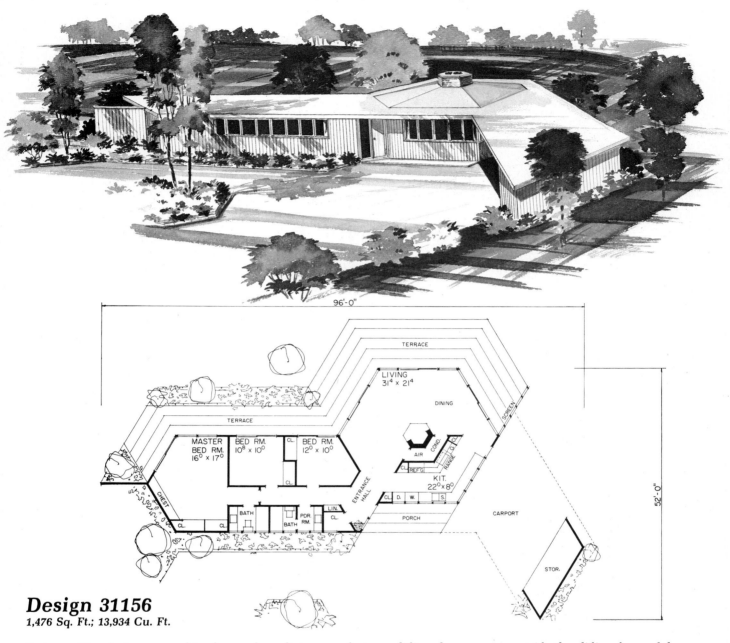

Design 31156
1,476 Sq. Ft.; 13,934 Cu. Ft.

● An exciting design, unusual in character, yet fun to live in. This frame home with its vertical siding and large glass areas has as its dramatic focal point a hexagonal living area which gives way to interesting angles. The large living area features sliding glass doors through which traffic may pass to the terrace stretching across the entire length of the house. The wide overhanging roof projects over the terrace and results in a large covered area outside the sliding doors of the master bedroom. The sloping ceilings converge above the unique open fireplace with its copper hood. The drive approach and the front entrance make this an eye-catching design.

Design 32591 1,428 Sq. Ft.; 21,725 Cu. Ft.

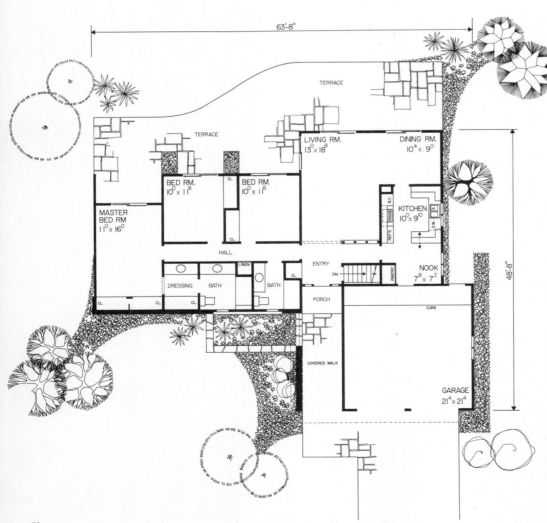

● A flowing terrace! The point of unity between interior and exterior in this distinctive home. This unusual terrace is accessible from every room except the kitchen . . . but designed to provide privacy as well. Inside, the spacious living and dining rooms feature two sets of sliding glass doors onto the terrace . . . allowing parties to spill outside during warm weather. And offering a scenic view all year round. Three bedrooms, all with sliding doors onto the terrace. Including a master suite with a dressing room and private bath. Plus its own secluded section of the terrace . . . perfect for solitary sunbathing or romantic nightcaps. Good times and easy work! There's an efficient kitchen with lots of work space and a large storage pantry. Plus a separate breakfast nook to make casual meals convenient and pleasant. This home creates its own peaceful enviroment! It's especially pleasing to people who love the outdoors.

Traditional
One-Story Homes
1600 - 2000 Sq. Ft.

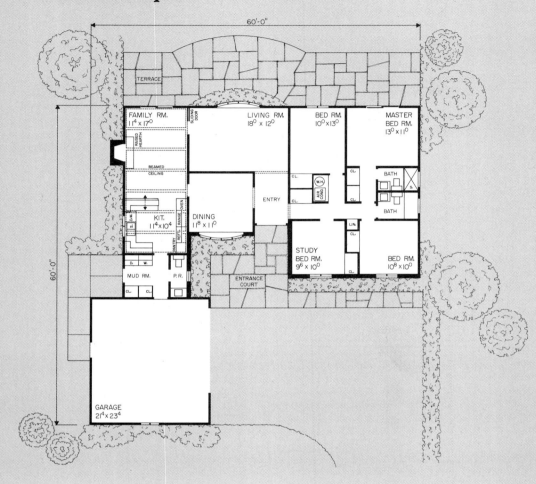

Design 32170
1,646 Sq. Ft.; 22,034 Cu. Ft.

● An L-shaped home with an enchanting Olde English styling. The wavy-edged siding, the simulated beams, the diamond lite windows, the unusual brick pattern and the interesting roof lines all are elements which set the character of authenticity. The center entry routes traffic directly to the formal living and sleeping zones of the house. Between the kitchen-family room area and the attached two-car garage is the mud room. Here is the washer and dryer with the extra powder room nearby. The family room is highlighted by the beamed ceilings, the raised hearth fireplace and sliding glass doors to the rear terrace. The work center with its abundance of cupboard space will be fun in which to function. Four bedrooms, two full baths and good closet space are features of the sleeping area.

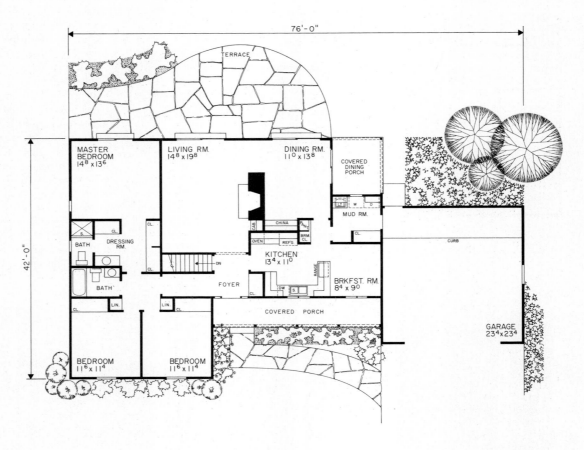

Design 32672 1,717 Sq. Ft.; 37,167 Cu. Ft.

● The traditional appearance of this one-story is emphasized by its covered porch, multi-paned windows, narrow clapboard and vertical wood siding. Not only is the exterior eye-appealing but the interior has an efficient plan and is very livable. The front U-shaped kitchen will work with the breakfast room and mud room, which houses the laundry facilities. An access to the garage is here. Outdoor dining can be enjoyed on the covered porch adjacent to the dining room. Both of these areas, the porch and dining room, are convenient to the kitchen. Sleeping facilities consist of three bedrooms and two full baths. Note the three sets of sliding glass doors leading to the terrace.

Design 31222 1,657 Sq. Ft.; 32,444 Cu. Ft.

● How will you call upon this home to function? As a three or four bedroom home? The study permits all kinds of flexibility in living patterns. If you wish, the extra room could serve the family as a TV area or an area for sewing, hobbies or guests. With two full baths plus an extra wash room, there is no lack of toilet facilities. The 27 foot living room with raised hearth fireplace features a formal dining area which is but a step from the secluded covered dining porch and efficient kitchen. There is a strategically located mud room which houses the washer and dryer and adjacent wash room. The stairs leading to the basement are also in this area.

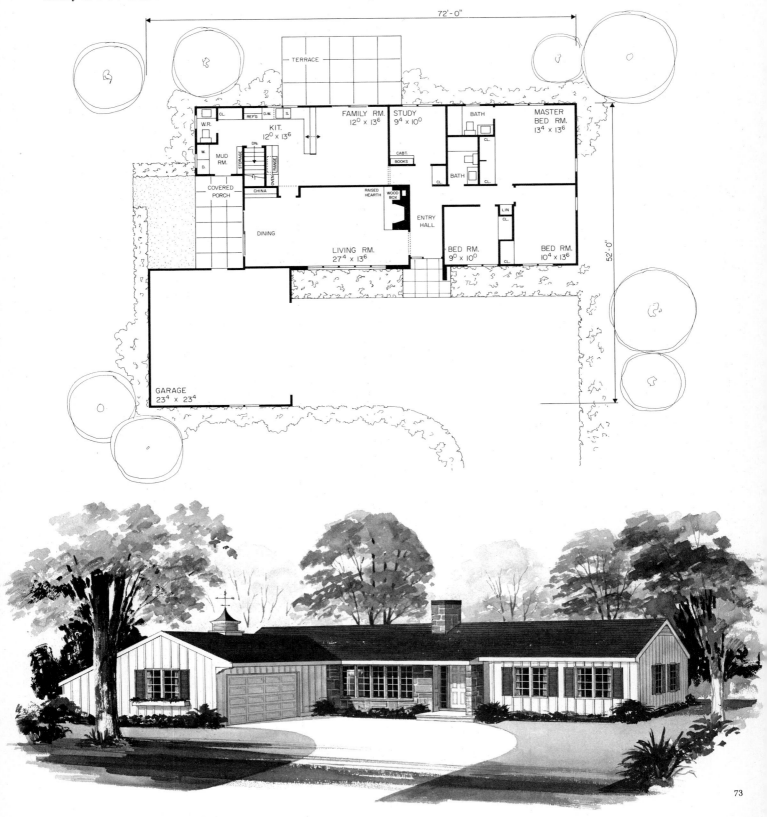

Design 31797
1,618 Sq. Ft.; 18,600 Cu. Ft.

● A house to be looked at and lived in —that's what this impressively formal French Provincial adaptation represents. The front court, just inside the brick wall with its attractive iron gate, sets the patterns of formality that are so apparent inside. The formal living and dining rooms separate the sleeping area from the kitchen/family room area. A pass-thru facilitates serving of informal snacks in family room. The three bedrooms will serve the average sized family perfectly. Details for an optional basement are included with this plan.

Design 31272
1,690 Sq. Ft.; 32,588 Cu. Ft.

● Designed for the family who wants the refinement of French Provincial, but on a small scale. Keynoting its charm are the long shutters, the delightful entrance porch with its wood posts, the interesting angles of the hip roof and the pair of paneled garage doors. Behind this formal facade is a simple, efficient, up-to-date floor plan. The living patterns for the family occupying this house will be toward the rear where the long living terrace is just a step outside the glass sliding doors of the formal and informal living areas. The two bathrooms in the sleeping area share back-to-back plumbing, which is economical.

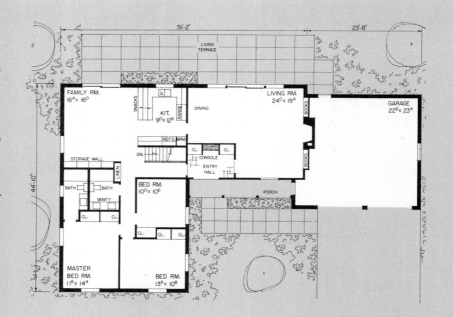

Design 31362
1,896 Sq. Ft.; 21,262 Cu. Ft.

● This dramatic L-shaped French adaptation is distinctive indeed. The projecting bedroom wing, consisting of three bedrooms, adds to the exterior appeal. Inside, the traffic patterns are excellent. The blueprints show optional basement details as well as an optional fireplace in the living room. This is most definitely a design to please all family members.

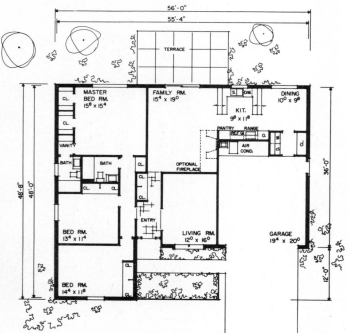

Design 31330
1,820 Sq. Ft.; 22,207 Cu. Ft.

● A dramatic, formal exterior using strong horizontal cornice lines to accentuate its low-slung qualities. The brick quoins at the corners, the paneled shutters, the ornamental iron and the hip-roof contribute to a distinctly continental flavor. Inside, the various areas are clearly defined. The bedrooms and living room are away from the noise of the family and service areas. The living room can double as a den-study. The U-shaped kitchen with pass-thru counters serves both the family room and dining area conveniently. The sleeping wing is highlighted by three sizable bedrooms, two baths and an abundance of wardrobe storage facilities. Optional fireplace details are included with the plan.

Design 31326
2,014 Sq. Ft.; 36,948 Cu. Ft.

● A pleasing French Provincial adaptation, with an extremely practical basic floor plan which contains many of those "extras" that contribute so much to good living. Among the features that make this such an outstanding home are: two fireplaces—one with raised hearth and wood box, first floor laundry, extra wash room, powder room, snack bar, separate dining room, partial basement, garage storage and work bench. With the four bedrooms and the formal and informal living areas, there is plenty of space in which the large family may move around and enjoy their surroundings.

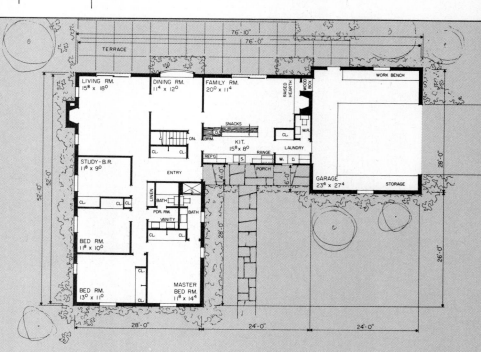

Design 31892
2,036 Sq. Ft.; 26,575 Cu. Ft.

● The romance of French Provincial is captured here by the hip-roof masses, the charm of the window detailing, the brick quoins at the corners, the delicate dentil work at the cornices, the massive centered chimney and the recessed double front doors. The slightly raised entry court completes the picture. The basic floor plan is a favorite of many. And little wonder, for all areas work well together, while still maintaining a fine degree of separation of functions. The highlight of the interior, perhaps, will be the sunken living room. The family room with its beamed ceiling will not be far behind in its popularity. The separate dining room, mud room and efficient kitchen complete the livability.

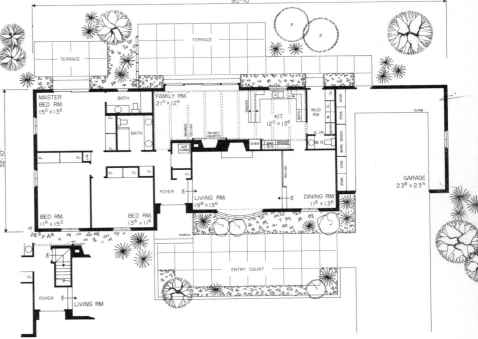

Three Distinctively Styled Exteriors . . .

Design 32705 1,746 Sq. Ft.; 37,000 Cu. Ft.

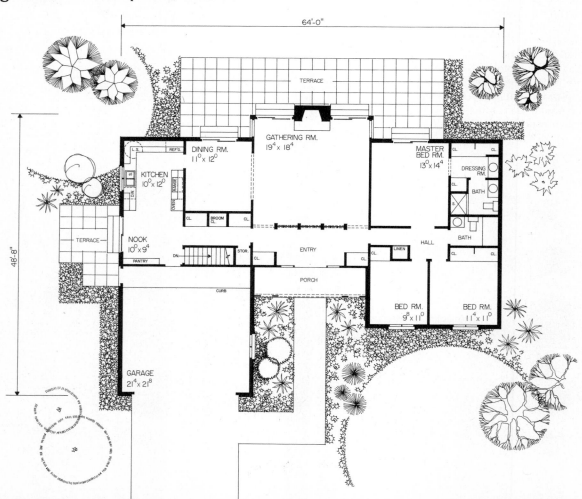

Design 32706 1,746 Sq. Ft.; 36,800 Cu. Ft.

... One Practical, Efficient Floor Plan

● Three different exteriors! But inside it's all the same livable house. Begin with the impressive entry hall ... more than 19' long and offering double entry to the gathering room. Now the gathering room which is notable for its size and design. Notice how the fireplace is flanked by sliding glass doors leading to the terrace! That's unusual. There's a formal dining room, too! The right spot for special birthday dinners as well as supper parties for friends. And an efficient kitchen that makes meal preparation easy whatever the occasion. Look for a built-in range and oven here ... plus a bright dining nook with sliding doors to a second terrace. Three large bedrooms! All located to give family members the utmost privacy. Including a master suite with a private dressing room, bath and a sliding glass door opening onto the main terrace. For blueprints of the hip-roof French adaptation on the opposite page order 32705. For the Contemporary version order 32706. The Colonial order 32704.

Design 32704 1,746 Sq. Ft.; 38,000 Cu. Ft.

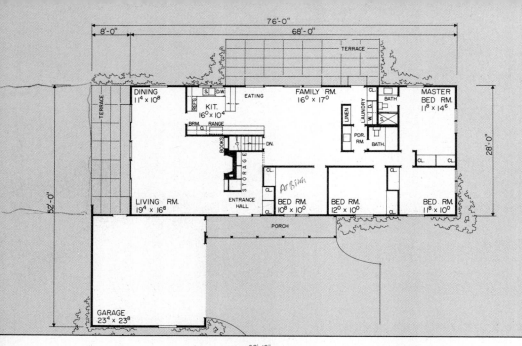

Design 31044
1,904 Sq. Ft.; 37,280 Cu. Ft.

● The impact of delightful design manifests itself in many different forms and numerous divergent styles. Here is a home that is a simple rectangle with a virtually square garage which go together to result in an interesting L-shape. The charming traditional styling is characterized by the window and door treatment, the horizontal and vertical siding, the roof masses and, above all, pleasing proportion. Inside there is a simple, straight-forward, plan.

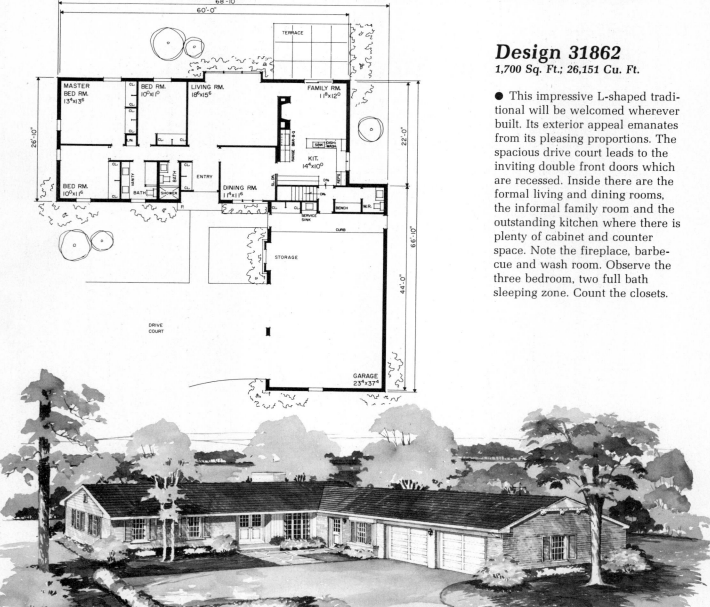

Design 31862
1,700 Sq. Ft.; 26,151 Cu. Ft.

● This impressive L-shaped traditional will be welcomed wherever built. Its exterior appeal emanates from its pleasing proportions. The spacious drive court leads to the inviting double front doors which are recessed. Inside there are the formal living and dining rooms, the informal family room and the outstanding kitchen where there is plenty of cabinet and counter space. Note the fireplace, barbecue and wash room. Observe the three bedroom, two full bath sleeping zone. Count the closets.

Design 31919
1,752 Sq. Ft.; 20,148 Cu. Ft.

● Here is a real charmer and one which your large family will have fun living in for years to come. All the facilities are present to assure convenient living. There are four good sized bedrooms. Two full back-to-back baths service the sleeping area efficiently and economically. There is the wonderfully spacious L-shaped living and dining area which looks out upon the rear terrace through sliding glass doors. The kitchen is ideally located between the dining and family rooms.

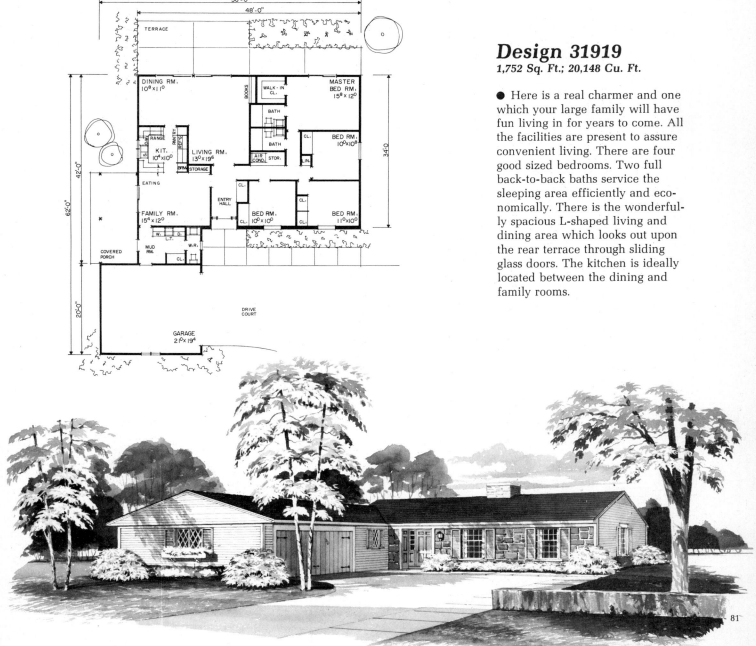

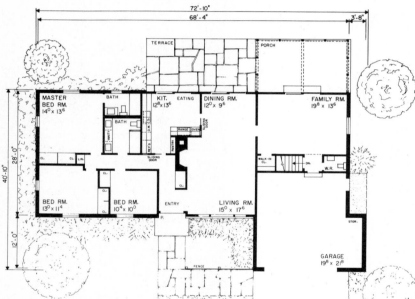

Design 31100
1,752 Sq. Ft.; 34,304 Cu. Ft.

● This modest sized, brick veneer home has a long list of things in its favor—from its appealing exterior to its feature-packed interior. All of the elements of its exterior complement each other to result in a symphony of attractive design features. The floor plan features three bedrooms, two full baths, an extra wash room, a family room, kitchen eating space, a formal dining area, two sets of sliding glass doors to the terrace and one set to the covered porch, built-in cooking equipment, a pantry and vanity with twin lavatories. Further, there is the living room fireplace, attached two-car garage with a bulk storage unit and a basement for extra storage and miscellaneous recreational activities. A fine investment.

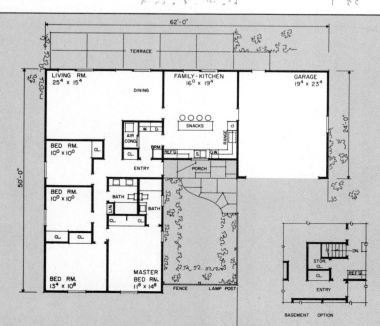

Design 31343
1,620 Sq. Ft.; 18,306 Cu. Ft.

● This is truly a prize-winner! The traditional, L-shaped exterior with its flower court and covered front porch is picturesque, indeed. The formal front entry routes traffic directly to the three distinctly zoned areas—the quiet sleeping area; the spacious; formal living and dining area; the efficient, informal family-kitchen. A closer look at the floor plan reveals four bedrooms, two full baths, good storage facilities, a fine snack bar and sliding glass doors to the rear terrace. The family-kitchen is ideally located. In addition to being but a few steps from both front and rear entrances, one will enjoy the view of both yards. Blueprints include basement and non-basement details.

Design 31896
1,690 Sq. Ft.; 19,435 Cu. Ft.

● Complete family livability is provided by this exceptional floor plan. Further, this design has a truly delightful traditional exterior. The fine layout features a center entrance hall with storage closet in addition to the wardrobe closet. Then, there is the formal, front living room and the adjacent, separate dining room. The U-shaped kitchen has plenty of counter and cupboard space. There is even a pantry. The family room functions with the kitchen and is but a step from the outdoor terrace. The mud room has space for storage and laundry equipment. The extra wash room is nearby. The large family will find those four bedrooms and two full baths just the answer to sleeping and bath accommodations.

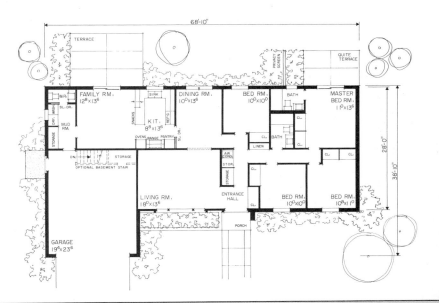

Design 31337
1,606 Sq. Ft.; 31,478 Cu. Ft.

● A pleasantly traditional facade which captures a full measure of warmth. Its exterior appeal results from a symphony of such features as: the attractive window detailing; the raised planter; the paneled door, carriage light and cupola of the garage; the use of both horizontal siding and brick. The floor plan has much to recommend this design to the family whose requirements include formal and informal living areas. There is an exceptional amount of livability in this modest-sized design.

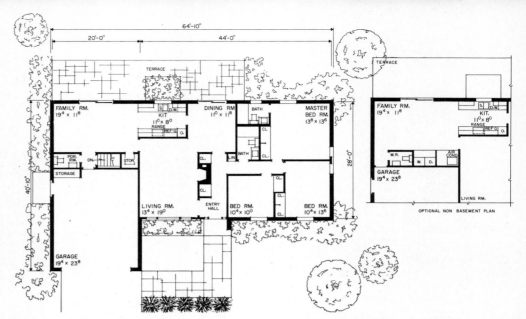

Design 31890
1,628 Sq. Ft.; 20,350 Cu. Ft.

● The pediment gable and columns help set the charm of this modestly sized home. Here is graciousness normally associated with homes twice its size. The pleasant symmetry of the windows and the double front doors complete the picture. Inside, each square foot is wisely planned to assure years of convenient living. There are three bedrooms, each with twin wardrobe closets. There are two full baths economically grouped with the laundry and heating equipment. A fine feature.

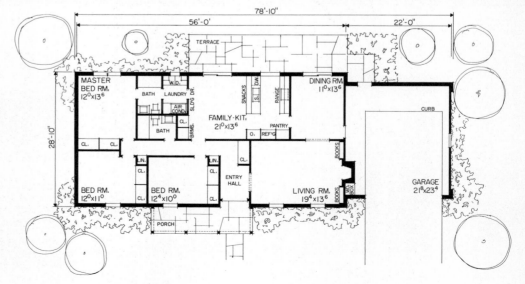

Design 31920
1,600 Sq. Ft.; 18,966 Cu. Ft.

● A charming exterior with a truly great floor plan. The front entrance with its covered porch seems to herald all the outstanding features to be found inside. Study the sleeping zone with its three bedrooms and two full baths. Each of the bedrooms has its own walk-in closet. Note the efficient U-shaped kitchen with the family and dining rooms to each side. Observe the laundry and the extra wash room. Blueprints for this design include details for both basement and non-basement construction.

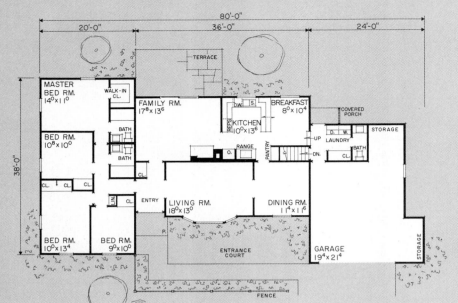

Design 31829
1,800 Sq. Ft.; 32,236 Cu. Ft.

● All the charm of a traditional heritage is wrapped up in this U-shaped home with its narrow, horizontal siding, delightful window treatment and high-pitched roof. The massive center chimney, the bay window and the double front doors are plus features. Inside, the living potential is outstanding. The sleeping wing is self-contained and has four bedrooms and two baths. The large family and living rooms cater to the divergent age groups.

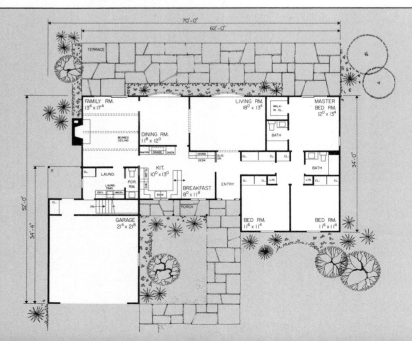

Design 31980
1,901 Sq. Ft.; 36,240 Cu. Ft.

● Planned for easy living, the daily living patterns of the active family will be pleasant ones, indeed. All the elements are present to assure a wonderful family life. The impressive exterior is enhanced by the recessed front entrance area with its covered porch. The center entry results in a convenient and efficient flow of traffic. A secondary entrance leads from the covered side porch, or the garage, into the first floor laundry. Note the powder room nearby.

Design 32360

1,936 Sq. Ft.; 37,026 Cu. Ft.

● There is no such thing as taking a fleeting glance at this charming home. Fine proportion and pleasing lines assure a long and rewarding study. Inside, the family's everyday routine will enjoy all the facilities which will surely guarantee pleasurable living. Note the sunken living room with its fireplace flanked by storage cabinets and book shelves. Observe the excellent kitchen just a step from the dining room and the nook.

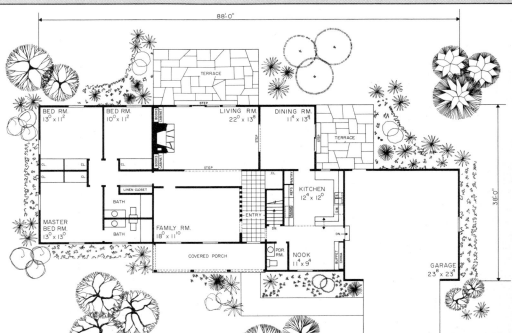

Design 32737
1,796 Sq. Ft.; 43,240 Cu. Ft.

● You will be able to build this distinctive, modified U-shaped one-story home on a relatively narrow site. But, then, if you so wished, with the help of your architect and builder you may want to locate the garage to the side of the house. Inside, the living potential is just great. The interior U-shaped kitchen handily services the dining and family rooms and nook. A rear covered porch functions ideally with the family room while the formal living room has its own terrace. Three bedrooms and two baths highlight the sleeping zone (or make it two bedrooms and a study). Notice the strategic location of the wash room, laundry, two storage closets and the basement stairs.

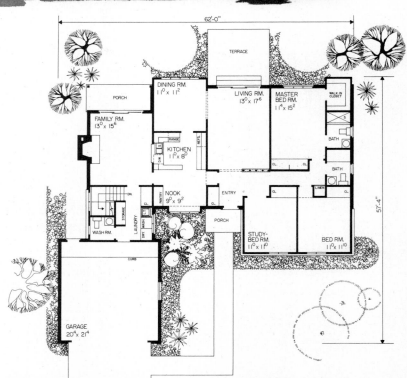

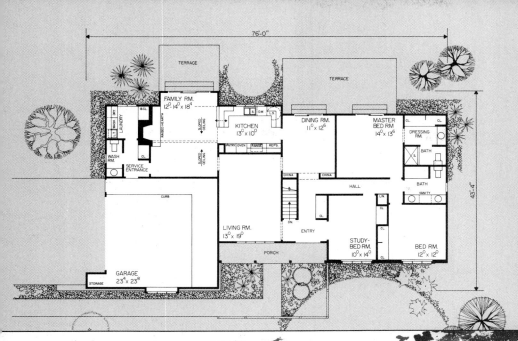

Design 32742
1,907 Sq. Ft.; 38,950 Cu. Ft.

● Colonial charm is expressed in this one-story design by the vertical siding, the post pillars, the cross fence, paned glass windows and the use of stone. A 19' wide living room, a sloped ceilinged family room with a raised hearth fireplace and its own terrace, a kitchen with many built-ins and a dining room with built-in china cabinets are just some of the highlights. The living terrace is accessible from the dining room and master bedroom. There are two more bedrooms and a full bath in addition to the master bedroom.

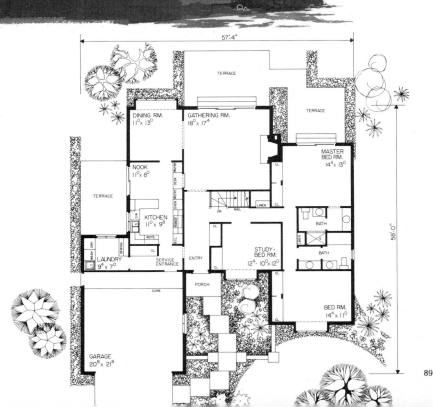

Design 32738
1,898 Sq. Ft.; 36,140 Cu. Ft.

● Impressive architectural work is indeed apparent in this three bedroom home. The three foot high entrance court wall, the high pitched roof and the paned glass windows all add to this home's appeal. It is also apparent that the floor plan is very efficient with the side U-shaped kitchen and nook with two pantry closets, the rear dining and gathering rooms and the three (or make it two with a study) bedrooms and two baths of the sleeping wing. Indoor-outdoor living also will be enjoyed in this home with a dining terrace off the nook and a living terrace off the gathering room and master bedroom. Note the fireplace in the gathering room and bay window in dining room.

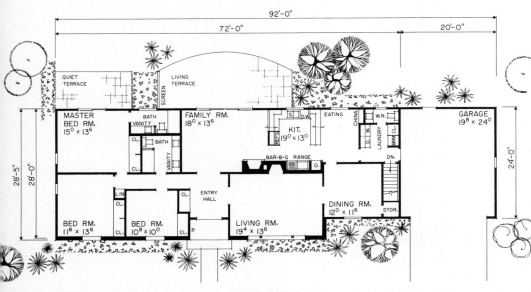

Design 31325
1,942 Sq. Ft.; 35,384 Cu. Ft.

● The large front entry hall permits direct access to the formal living room, the sleeping area and the informal family room. Both of the living areas have a fireplace. When formal dining is the occasion of the evening the separate dining room is but a step from the living room. The U-shaped kitchen is strategically flanked by the family room and the breakfast areas.

Design 32316
2,000 Sq. Ft.; 25,242 Cu. Ft.

● If you are looking for a four bedroom version of the two other designs on this page, look no further. Here, in essentially the same number of square feet, is a Colonial adaptation for a larger family. The floor planning of this basic design results in excellent zoning. The four bedroom, two-bath sleeping zone comprises a wing of its own directly accessible from the main foyer.

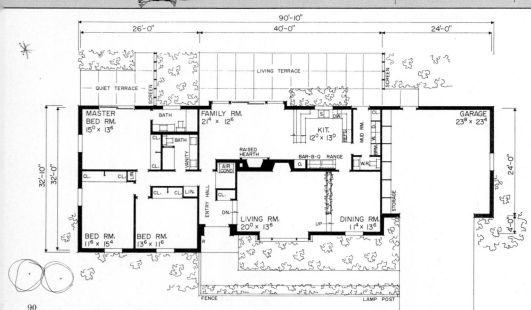

Design 31748
1,986 Sq. Ft.; 23,311 Cu. Ft.

● A sunken living room, two fireplaces, 2½ baths, a rear family room, a formal dining room, a mud room and plenty of storage facilities are among the features of this popular design. Blueprints include optional basement details.

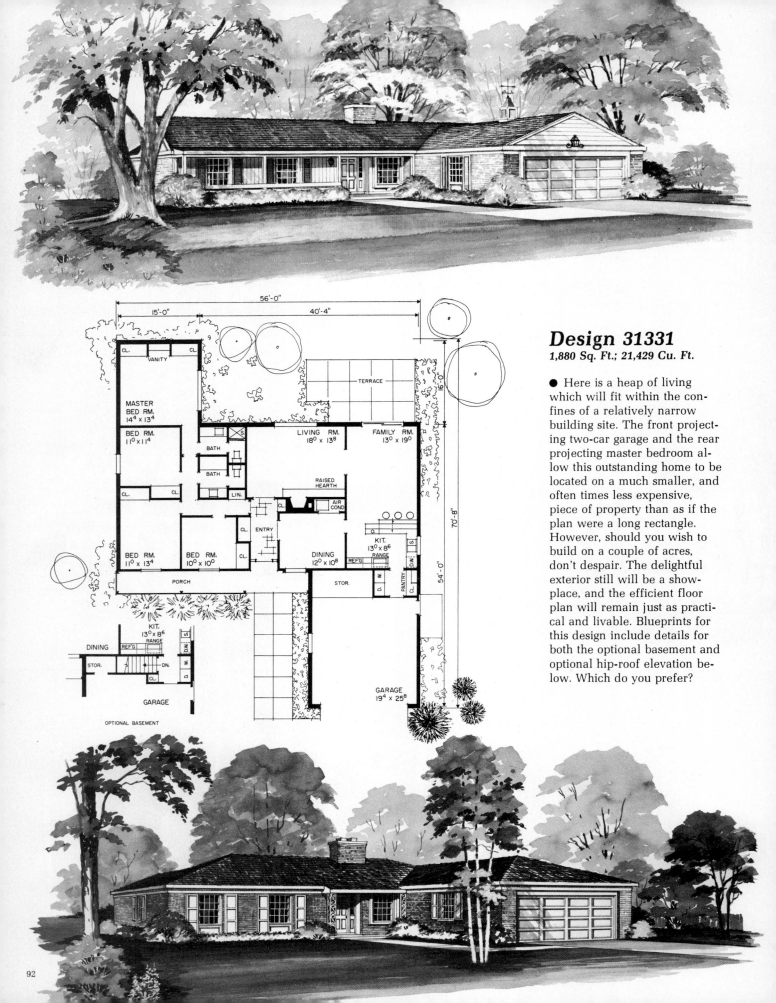

Design 31331
1,880 Sq. Ft.; 21,429 Cu. Ft.

● Here is a heap of living which will fit within the confines of a relatively narrow building site. The front projecting two-car garage and the rear projecting master bedroom allow this outstanding home to be located on a much smaller, and often times less expensive, piece of property than as if the plan were a long rectangle. However, should you wish to build on a couple of acres, don't despair. The delightful exterior still will be a showplace, and the efficient floor plan will remain just as practical and livable. Blueprints for this design include details for both the optional basement and optional hip-roof elevation below. Which do you prefer?

Design 31317   1,930 Sq. Ft.; 23,573 Cu. Ft.

● The paneled front door, flanked by attractive vertical glass panels, of the recessed entrance will welcome callers to this charming traditional exterior with a refreshingly different floor plan. The highlight of the interior is the unique location of the family room. While it functions conveniently with the kitchen, it is also easily accessible from the formal front entry hall, the attached two-car garage and the covered rear porch. A raised hearth fireplace, a snack bar and sliding glass doors are plus features of this multi-purpose area. The strategic location of the extra wash room will reduce thru-the-house traffic from the outdoors. The spaciousness, resulting from the open planning of the formal living and dining rooms, permits the creation of a most gracious atmosphere. Observe the built-in bookcases adjacent to the living room and entry hall.

Design 31130
1,856 Sq. Ft.; 36,044 Cu. Ft.

● A delightfully impressive and formal ranch home. The brick veneer, the ornamental iron columns supporting the overhanging gable, the window treatment and the cupola are among the features that contribute to the charm of this almost perfectly rectangular home. A study of the floor plan reveals excellent room relationships. The formal areas, the living and dining rooms, are adjacent to each other and overlook the front yard. The efficient kitchen is flanked by the informal breakfast area and the all-purpose family room.

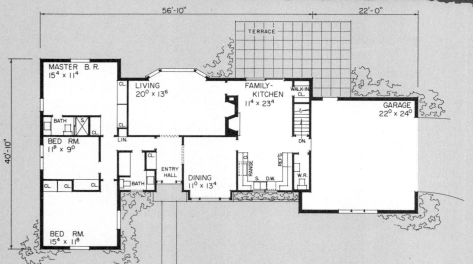

Design 31091
1,666 Sq. Ft.; 28,853 Cu. Ft.

● What could be finer than to live in a delightfully designed home with all the charm of the exterior carried right inside. The interior points of interest are many. However, the focal point will surely be the family-kitchen. The work center is U-shaped and most efficient. The family activity portion of the kitchen features an attractive fireplace which will contribute to a feeling of warmth and fellowship. Nearby is the wash room and stairs to the basement.

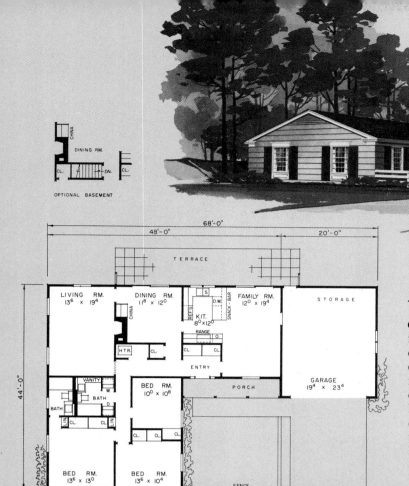

Design 31385   1,632 Sq. Ft.; 19,384 Cu. Ft.

● The charm of this colonial home is to be found in its appealing proportion and its outstanding window details. The wood fence with its lamp post helps create an inviting court adjacent to the covered front entrance. A study of the floor plan reveals excellent zoning. The efficient kitchen functions ideally with the family room which opens onto the outdoor terrace through sliding glass doors. Conveniently located between the quiet living room and the kitchen is the separate dining room. It, too, opens onto the terrace. Blueprints for this design include basement details.

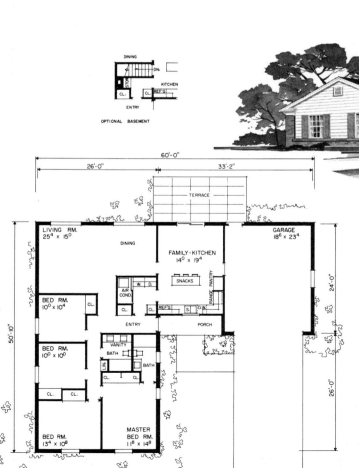

Design 31384   1,636 Sq. Ft.; 18,536 Cu. Ft.

● Here, a wise and economical use of space results in a most practical plan with exceptional livability. There are four bedrooms, two full baths, a 25-foot formal living and dining area, a wonderful family kitchen, an attached two-car garage and a covered front porch. For those who wish to build with a basement there is the optional plan included with the blueprints. Features that are particularly noteworthy are the private bath for the master bedroom, the twin lavatories for the main bath, the island snack bar of the kitchen, the extra closets and the pantry in the kitchen.

Design 32605
1,775 Sq. Ft.; 34,738 Cu. Ft.

● Here are three modified L-shaped Tudor designs with tremendous exterior appeal and efficient floor plans. While each plan features three bedrooms and 2½ baths, the square footage differences are interesting. Note that each design may be built with or without a basement. This appealing exterior is highlighted by a variety of roof planes, patterned brick, wavy-edged siding and a massive chimney. The garage is oversized and has good storage potential. In addition to the entrance court, there are two covered porches and two terraces for outdoor living. Most definitely a home to be enjoyed by all family members.

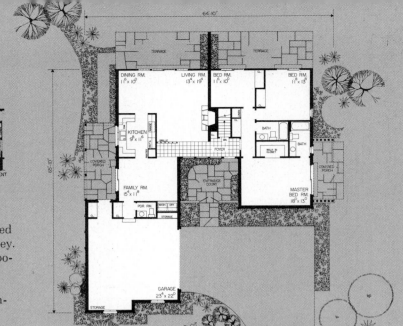

Design 32206
1,769 Sq. Ft.; 25,363 Cu. Ft.

● The charm of Tudor adaptations has become increasingly popular in recent years. And little wonder. Its freshness of character adds a unique touch to any neighborhood. This interesting one-story home will be a standout wherever you choose to have it built. The covered front porch leads to the formal front entry–the foyer. From this point traffic flows freely to the living and sleeping areas. The outstanding plan features a separate dining room, a beamed ceiling living room, an efficient kitchen and an informal family room.

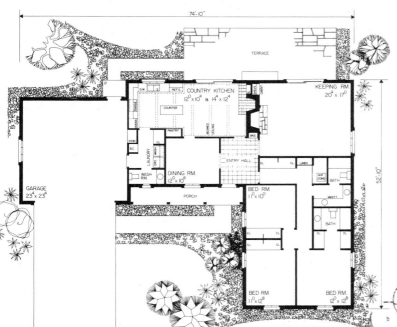

Design 32604
1,956 Sq. Ft.; 28,212 Cu. Ft.

● A feature that will set the whole wonderful pattern of true living will be the 26 foot wide country kitchen. The spacious, L-shaped kitchen has its efficiency enhanced by the island counter work surface. Beamed ceilings, fireplace and sliding glass doors add to the cozy atmosphere of this area. The laundry, dining room and entry hall are but a step or two away. The big keeping room also has a fireplace and can function with the terrace. Observe the 2½ baths.

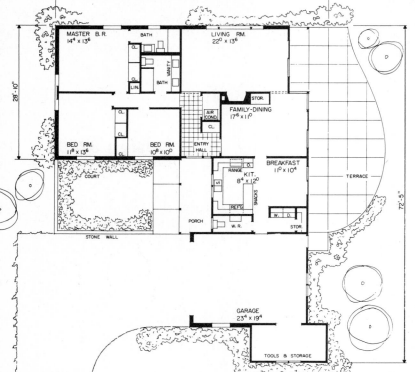

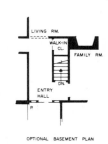

Design 33144
1,760 Sq. Ft.; 19,240 Cu. Ft.

● If you are short on space and searching for a home that is long on both good looks and livability, search no more! This impressive L-shaped home measures merely 56'-5" in width. It, therefore, qualifies for placement on a relatively narrow building site. Of course with land costs so high, the purchase of a smaller and less expensive building site can significantly reduce the building budget. Whether you build with, or without, a basement (blueprints include details for both types of construction) the outstanding livability remains. There are three bedrooms, two baths, a formal rear living room, a big breakfast room, an excellent kitchen, a wash room and laundry and a huge bulk storage area projecting from the front of the garage. Note window treatment.

98

Design 31280
1,730 Sq. Ft.; 33,873 Cu. Ft.

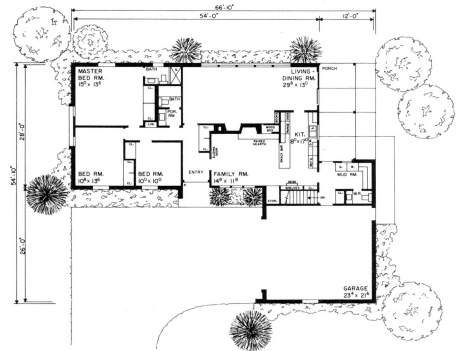

● This medium-sized, L-shaped ranch home with its touch of traditional styling is, indeed, pleasing to the eye. Hub of this plan is the centrally located foyer which controls traffic to the bedrooms, living/dining room and family room. Economically planned circulation spends very little floor space on hallways. Back-to-back fireplaces in the family room and the living room are an effective sound barrier between the informal and formal areas. Strategically placed kitchen allows for a convenient observation post overlooking family room. The kitchen features a snack bar, built-in desk with shelves and adjacent mud room with washer, dryer and wash room. This design includes a basement for future development or storage space.

Design 32261
1,825 Sq. Ft.; 33,814 Cu. Ft.

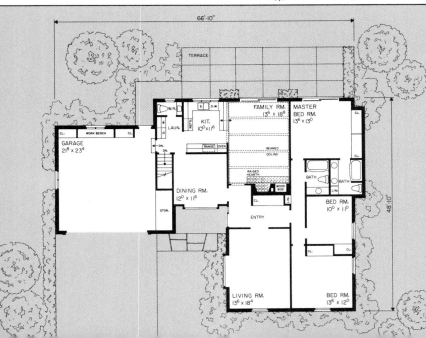

● This distinctive L-shaped home virtually exudes traditional warmth and charm. And little wonder, for the architectural detailing is, indeed, exquiste. Notice the fine window detailing, the appealing cornice work, the attractiveness of the garage door and the massive chimney. The dovecote and the weathervane add to the design impact. The covered front porch shelters the entry which is strategically located to provide excellent traffic patterns. A service entry from the garage is conveniently located handy to the laundry, wash room, kitchen and stairs to the basement. The beamed ceilinged family room will naturally be everyone's favorite spot for family living.

Design 32603
1,949 Sq. Ft.; 41,128 Cu. Ft.

● Surely it would be difficult to beat the appeal of this traditional one-story home. Its slightly modified U-shape with the two front facing gables, the bay window, the covered front porch and the interesting use of exterior materials all add to the exterior charm. Besides, there are three large bedrooms serviced by two full baths and three walk-in closets. The excellent kitchen is flanked by the formal dining room and the informal family room. Don't miss the pantry, the built-in oven and the pass-thru to the snack bar. The handy first floor laundry is strategically located to act as a mud room. The extra wash room is but a few steps away. The sizable living room highlights a fireplace and a picture window. Note the location of the basement stairs.

Design 33177
1,888 Sq. Ft.; 21,264 Cu. Ft.

● It would certainly be difficult to pack more living potential into such a modestly sized home than this. The family's living patterns will be just great with "convenience" as the byword. Consider: three bedrooms serviced by two full baths; a formal living room and a formal dining room overlooking the rear terrace and free from unnecessary cross-room traffic; an interior kitchen functioning ideally with the dining room, the family room and the laundry; a powder room handy to living and family rooms and kitchen. Observe: stall shower, fireplace, abundant wardrobe closets, built-in planter, snack bar, family room storage wall and sliding glass doors. The exterior is charming.

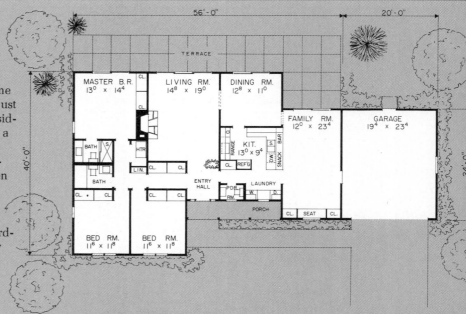

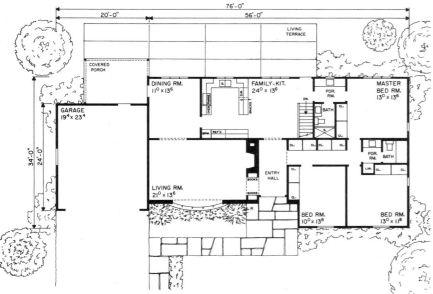

Design 31282
1,732 Sq. Ft.; 33,913 Cu. Ft.

● If it is just good plain economical livability with an exterior that has a flair for distinction that you are seeking, you'll find this traditional home difficult to top. Within only 1,732 square feet there are all the features the average family could require, plus a few for good measure. Observe that there are two full baths with one compartmented in such a manner as to conveniently serve two distinct areas—the master bedroom and the family-kitchen. There won't be any lack of storage facilities with all that cupboard and closet space. The 21 foot living room has a fireplace and built-in book shelves. There is the formal dining room and then, the covered porch nearby for warm evening relaxation. The family-kitchen is ideal.

Design 31252
1,985 Sq. Ft.; 24,924 Cu. ft.

● Here is a traditional adaptation which embodies all the warmth and appeal to make its occupants swell with pride. The projection of the bedroom wing, the recessed front entrance, the boxed bay windows and the set-back attached two-car garage all contribute to the interesting lines of the front exterior. The floor plan is one which contains all the features an active family would require to assure years of exceptional livability. There are two full baths in the three bedroom sleeping area, a sunken formal living room, a separate dining room and a spacious rear family room.

Design 32550
1,892 Sq. Ft.; 39,590 Cu. Ft.

● An enchanting low-slung traditional ranch with exceptional appeal. The low-pitched roof has a wide overhang and exposed beams. Stone and vertical siding offer a pleasing contrast. However, you may wish to substitute other materials of your choice. The diamond lite windows, the fence with its lamp post, the double front doors and the dovecote above the carriage lamp of the garage are among the interesting exterior features. Inside, there are four bedrooms and two full baths in the sleeping wing. The L-shaped living area is spacious and features a sloping ceiling for the gathering and dining rooms. The open stairwell to the basement recreation area is attractive. The pleasant kitchen is flanked by the nook and laundry.

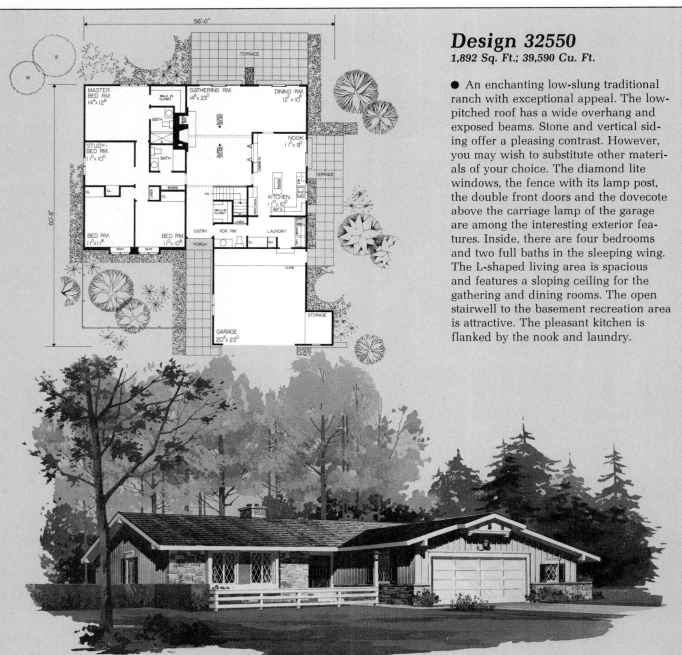

Design 31186
1,872 Sq. Ft.; 31,680 Cu. Ft.

● This appealing home has an interesting and practical floor plan. It is cleverly zoned to cater to the living patterns of both the children and the parents. The children's bedroom wing projects to the rear and functions with their informal family room. The master bedroom is ideally isolated and is located in a part of the plan's quietly formal wing. The efficient kitchen looks out upon the rear terrace and functions conveniently with the dining area and family room. A full bath serves each of the two main living areas. A built-in vanity highlights each bath. The mud room features laundry equipment, storage unit and stairs to the basement. The blueprints show details for basement and non-basement construction.

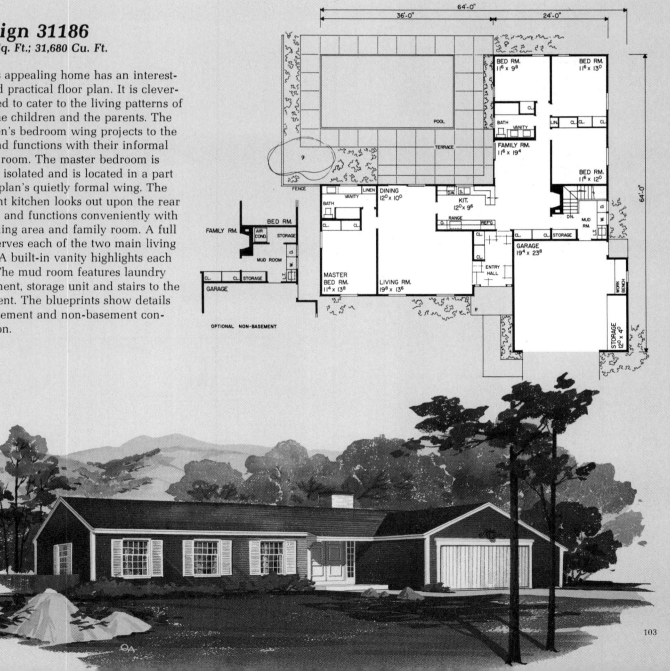

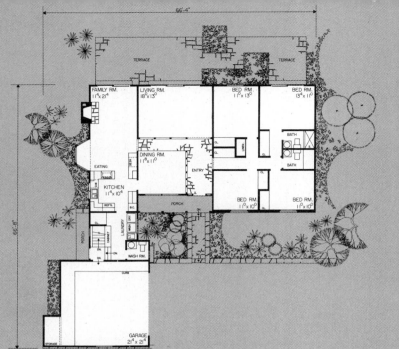

Design 32533
1,897 Sq. Ft.; 40,523 Cu. Ft.

● The distinctive appeal of the traditional, L-shaped ranch home is indeed, hard to beat. Particularly, one with such exquisite exterior appointments. Notice the delightful window and door treatment, the covered front porch, the vertical siding and the fieldstone, the dovecote and the carriage lamp. The center entrance with its slate floor routes traffic effectively to all areas. The four bedroom sleeping wing highlights two full baths. The formal living and dining rooms act as a buffer between sleeping area and the all-purpose family room/kitchen zone. The family room has sliding glass doors, a fireplace and a large bay window for extra informal eating space. There is a first floor laundry, an extra wash room and a basement.

Design 32677
1,634 Sq. Ft.; 26,770 Cu. Ft.

● Showing Spanish influence by utilizing a stucco exterior finish, grilled windows and an arched entryway, this one-story design is just what you've been waiting for. Beyond the arched entryway is the private front court which leads to the tiled foyer. Interior livability has been well planned. The rear living room is warmed by a fireplace and has sliding glass doors to the terrace as does the informal family room. The U-shaped kitchen will easily serve the dining room. This room has access to the front court for outdoor dining. Three bedrooms are in the sleeping wing; the master bedroom has a private bath, dressing room with walk-in closet and sliding glass doors leading to a private covered porch.

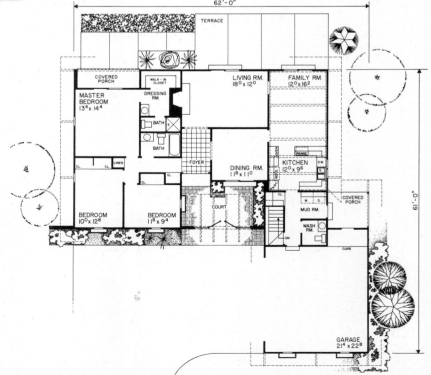

Design 32678
1,971 Sq. Ft.; 42,896 Cu. Ft.

● If you've ever desired to have a large country kitchen in your home then this is the design for you. The features of this room are many, indeed. Begin your list with the island range with snack bar, pantry and broom closets, eating area with sliding glass doors leading to a covered porch, adjacent mud room with laundry facilities, raised hearth fireplace and conversation area with built-in desk on one side and shelves on the other. Now that is some multi-purpose room! Review the rest of this plan which is surrounded by a delightful Tudor facade. It will surely prove to be remarkable.

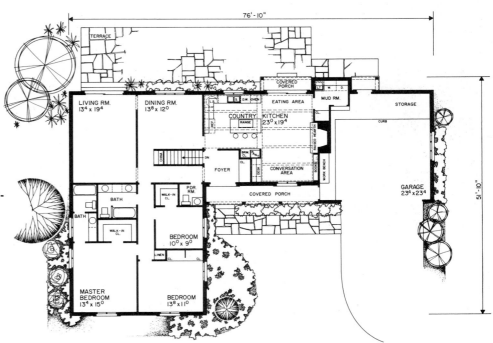

Design 32277

1,903 Sq. Ft.; 25,087 Cu. Ft.

● Tudor design front and center! And what an impact this beautifully proportioned L-shaped home does deliver. Observe the numerous little design features which make this such an attractive home. The half-timber work, the window styling, the front door detailing, the covered porch post brackets and the chimney are all among the delightful highlights. Well-zoned, the dining and living rooms are openly planned for formal dining and living.

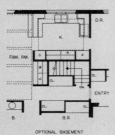

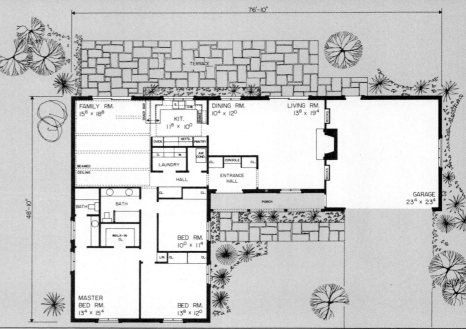

Design 32728

1,825 Sq. Ft.; 38,770 Cu. Ft.

● Your family's new lifestyle will surely flourish in this charming, L-shaped English adaptation. The curving front driveway produces an impressive approach. A covered front porch shelters the centered entry hall which effectively routes traffic to all areas. The fireplace is the focal point of the spacious, formal living and dining area. The kitchen is strategically placed to service the dining room and any informal eating space developed in the family room. In addition to the two full baths of the sleeping area, there is a handy wash room at the entrance from the garage. A complete, first floor laundry is nearby and has direct access to the yard. Sliding glass doors permit easy movement to the outdoor terrace and side porch. Don't overlook the basement and its potential for the development of additional livability and/or storage.

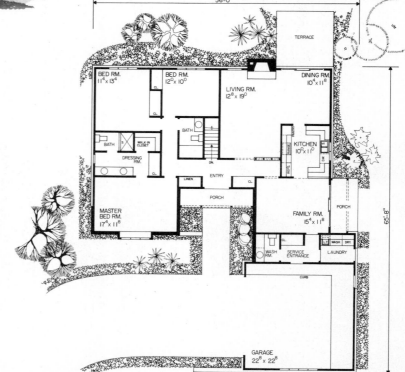

Design 32374
1,919 Sq. Ft.; 39,542 Cu. Ft.

● This English adaptation will never grow old. There is, indeed, much here to please the eye for many a year to come. The wavy-edged siding contrasts pleasingly with the diagonal pattern of brick below. The diamond lites of the windows create their own special effect. The projecting brick wall creates a pleasant court outside the covered front porch. The floor plan is well-zoned with the three bedrooms and two baths comprising a distinct sleeping wing. Flanking the entrance hall is the formal living room and the informal, multi-purpose family room. The large dining room is strategically located. The mud room area is adjacent to the extra wash room and the stairs to the basement.

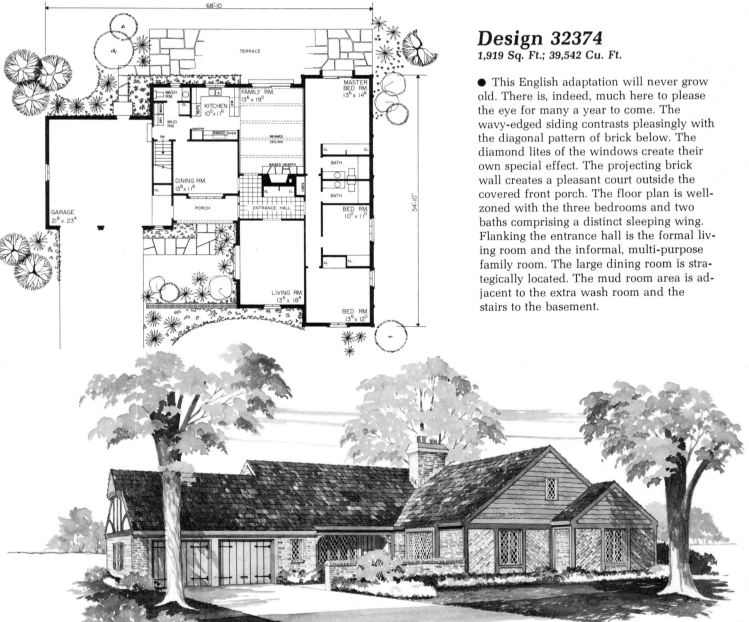

Design 31758
1,872 Sq. Ft.; 20,878 Cu. Ft.

● Setting the delightful character of this L-shaped traditional one-story home are such design features as: the wood columns of the covered front porch; the pediment, gabled ends; the attractive window treatment and the front entrance detail. The floor plan is an exceptional one. The nicely sized formal entry hall routes traffic ideally to the major areas of the house. The two large living areas - the informal family-kitchen and the quiet, formal living room - look out upon the rear terrace through sliding glass doors. The work center is efficient.

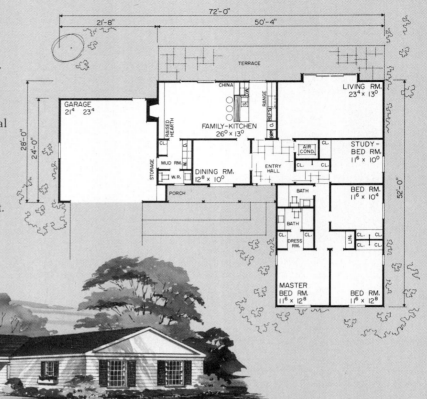

Design 31962
1,762 Sq. Ft.; 20,063 Cu. Ft.

● Handsome to look at, and just as appealing in which to live. Wide, horizontal siding, attractive window treatment, recessed double front doors, gabled ends, overhanging roof - these are the features which are so delightfully complimented by the rail fence with its gas light lamp post. This home will be ideal for any size family. The front entry is strategically located to route traffic directly to each of the main areas of the house.

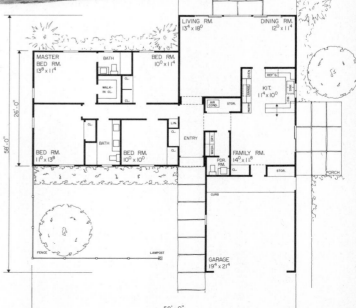

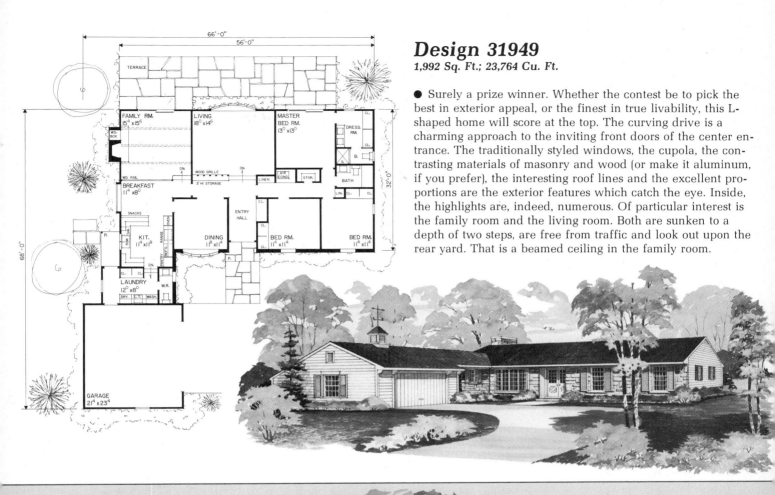

Design 31949
1,992 Sq. Ft.; 23,764 Cu. Ft.

● Surely a prize winner. Whether the contest be to pick the best in exterior appeal, or the finest in true livability, this L-shaped home will score at the top. The curving drive is a charming approach to the inviting front doors of the center entrance. The traditionally styled windows, the cupola, the contrasting materials of masonry and wood (or make it aluminum, if you prefer), the interesting roof lines and the excellent proportions are the exterior features which catch the eye. Inside, the highlights are, indeed, numerous. Of particular interest is the family room and the living room. Both are sunken to a depth of two steps, are free from traffic and look out upon the rear yard. That is a beamed ceiling in the family room.

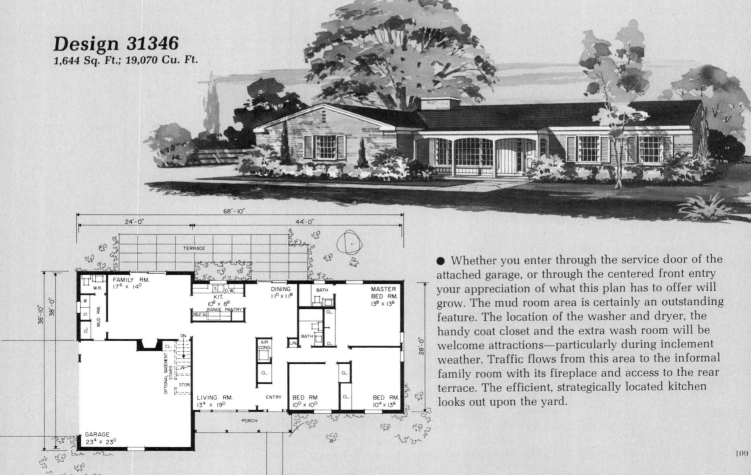

Design 31346
1,644 Sq. Ft.; 19,070 Cu. Ft.

● Whether you enter through the service door of the attached garage, or through the centered front entry your appreciation of what this plan has to offer will grow. The mud room area is certainly an outstanding feature. The location of the washer and dryer, the handy coat closet and the extra wash room will be welcome attractions—particularly during inclement weather. Traffic flows from this area to the informal family room with its fireplace and access to the rear terrace. The efficient, strategically located kitchen looks out upon the yard.

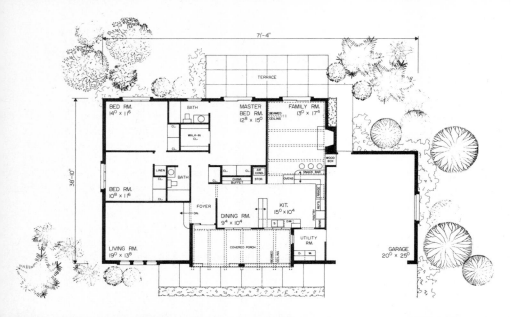

Design 32232
1,776 Sq. Ft.; 17,966 Cu. Ft.

● This appealing, flat roof design has its roots in the Spanish Southwest. The arched, covered porch with its heavy beamed ceiling sets the note of distinction. The center foyer routes traffic effectively to the main zones of the house. Down a step is the sunken living room. Privacy will be the byword here. The cluster of three bedrooms features two full baths and good storage facilities.

Design 32200
1,695 Sq. Ft.; 18,916 Cu. Ft.

● If you have a penchant for something delightfully different this Spanish adaptation may be just what you've been waiting for. This ranch home with an L-shape will go well on any site - large or small. A popular feature will be the entry court. Of great interest is the indoor-outdoor relationships. Observe how the major rooms function through sliding glass doors with the terraces.

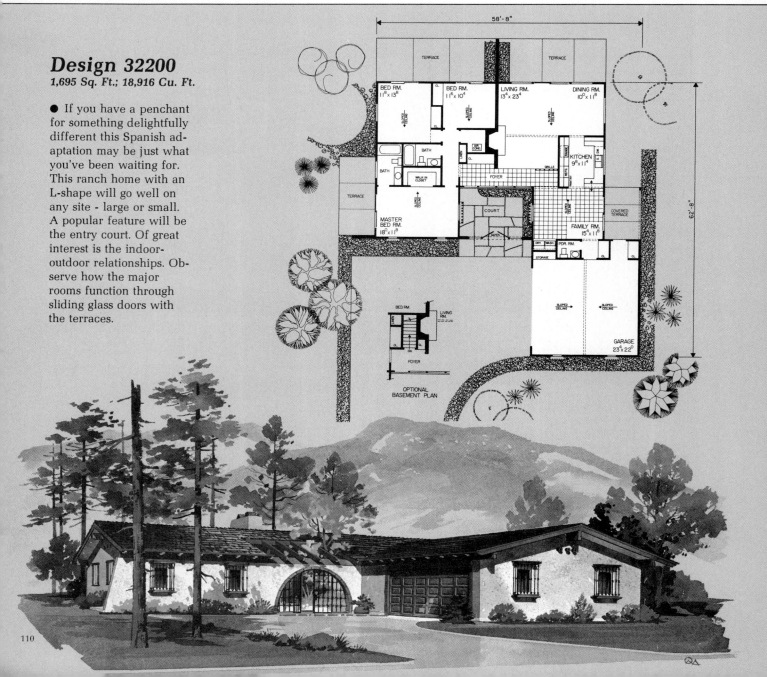

Design 31726
1,910 Sq. Ft.; 19,264 Cu. Ft.

● The U-shaped plan has long been honored for its excellent zoning. As the floor plan for this fine Spanish adaptation illustrates, it not only provides separation between parents' area and children's wing, but also it places a buffer area in the center. This makes the kitchen the "control center" for the home - handy to the family room, living room and the dining alcove.

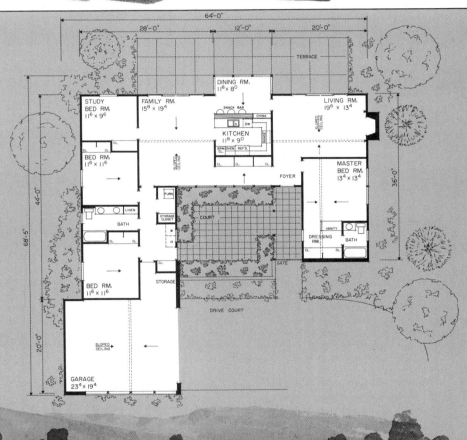

111

Design 32265 1,861 Sq. Ft.; 36,888 Cu. Ft.

● Inside and outside this home will provide a lifetime of satisfaction for its occupants. Whether called upon to serve the family's needs as a two or three bedroom home, it will do so with great merit. As a two bedroom home, there is then a quiet study with built-in desk, cabinets and book shelves. The front-to-rear living room is 23 feet long and functions through sliding glass doors right out into the terrace. Around the corner from the fireplace is the formal dining room. It, too, but a step from the terrace. The in-line kitchen will be efficient and function with breakfast nook. Note built-in china cabinets, pantry, wood box, vanity, linen storage and basement.

● Certainly a unique adaptation of an Early American farm. Projecting from the main portion of the house are the living and garage wings. Anyone wishing to build an expansible house would find this design of interest. The main section of the house comprised of the bedroom, kitchen and family room areas would function ideally as the initial basic unit. Later, as the need arose, the spacious living-dining wing could be added. Then the two-car garage could be built.

Design 32210 1,658 Sq. Ft.; 22,804 Cu. Ft.

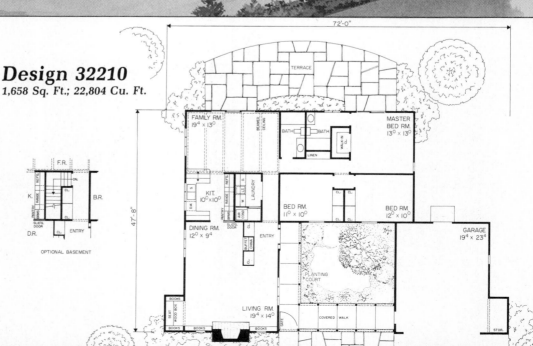

Design 32327
1,820 Sq. Ft.; 26,897 Cu. Ft.

● You will have fun deciding which orientation on your site you most prefer for this brick veneer home. Its angular shape provides just the right touch to assure an air of distinction. The center foyer routes traffic effectively. There is no unnecessary cross-room traffic in this design. The living room with access to the rear privacy porch features an appealing raised hearth fireplace wall. Acting as a divider between the nook and family room is a counter with cabinets above and below. A pass-thru provides direct access to each room. Should you wish, you may opt to have the family room function as the dining room. The built-in planter is an attractive highlight.

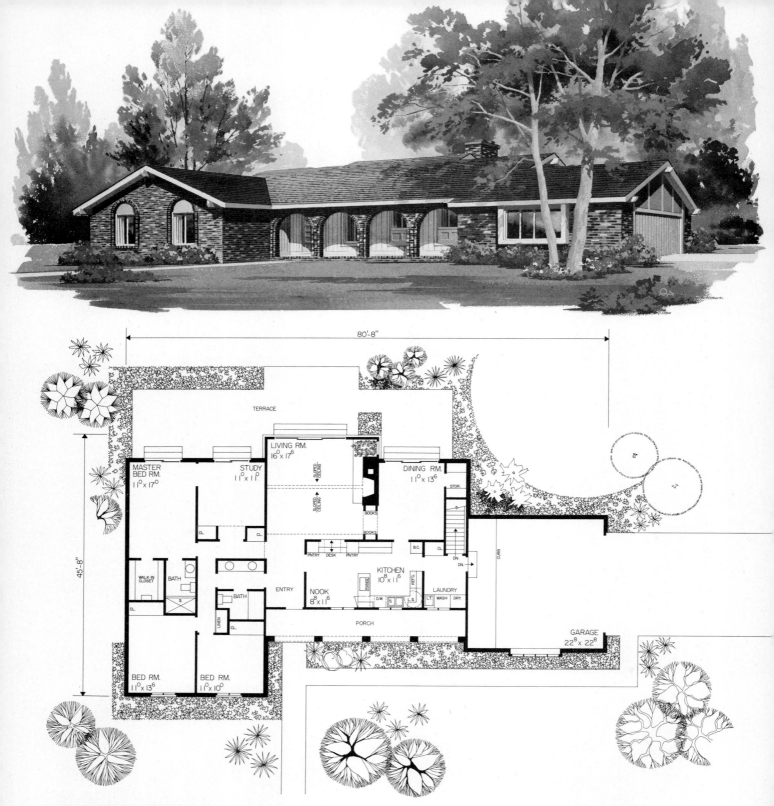

Design 32557 1,955 Sq. Ft.; 43,509 Cu. Ft.

● This eye-catching design with a flavor of the Spanish Southwest will be as interesting to live in as it will be to look at. The character of the exterior is set by the wide overhanging roof with its exposed beams; the massive arched pillars; the arching of the brick over the windows; the panelled door and the horizontal siding that contrasts with the brick. The master bedroom/study suite is a focal point of the interior. However, if necessary, the study could become the fourth bedroom. The living and dining rooms are large and are separated by a massive raised hearth fireplace. Don't miss the planter, the book niches and the china storage. The breakfast nook and the laundry flank the U-shaped kitchen. Notice the twin pantries, the built-in planning desk and the pass thru. That's a big lazy susan to the right of the kitchen sink. Notice the twin lavatories in the big main bath.

Contemporary
One-Story Homes
1600 - 2000 Sq. Ft.

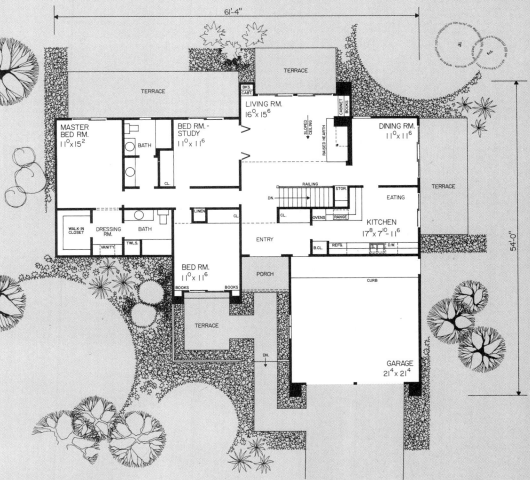

Design 32702
1,636 Sq. Ft.; 38,700 Cu. Ft.

● A rear living room with a sloping ceiling, built-in bookcases, a raised hearth fireplace and sliding glass doors to the rear living terrace. If desired, bi-fold doors permit this room to function with the adjacent study. Open railing next to the stairs to the basement recreation area fosters additional spaciousness. The kitchen has plenty of cabinet and cupboard space. It features informal eating space and is but a step or two from the separate dining room. Note side dining terrace. Each of the three rooms in the sleeping wing has direct access to outdoor living. The master bedroom highlights a huge walk-in wardrobe closet, dressing room with built-in vanity and private bath with large towel storage closet. Projecting the two-car garage with its twin doors to the front not only contributes to an interesting exterior, but reduces the size of the building site required for this home. A lot of living from 1,636 square feet.

Design 32741

1,842 Sq. Ft.; 37,045 Cu. Ft.

● Here is another example of what 1,800 square feet can deliver in comfort and convenience. The setting reminds one of the sun country of Arizona. However, this design would surely be an attractive and refreshing addition to any region. The covered front porch with its adjacent open trellis area shelters the center entry. From here traffic flows efficiently to the sleeping, living and kitchen zones. There is much to recommend each area. The sleeping with its fine bath and closet facilities; the living with its spaciousness, fireplace and adjacent dining room; the kitchen with its handy nook, excellent storage, nearby laundry and extra wash room.

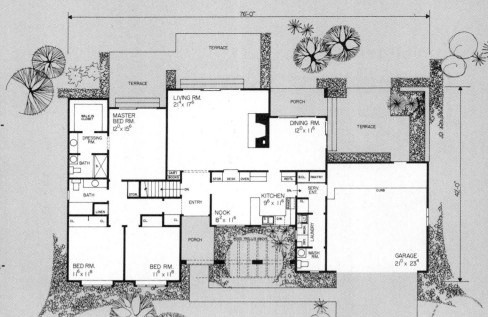

Design 32386

1,994 Sq. Ft.; 22,160 Cu. Ft.

● This distinctive home may look like the Far West, but don't let that inhibit you from enjoying the great livability it has to offer. Wherever built, you will surely experience a satisfying pride of ownership. Imagine, an entrance court in addition to a large side courtyard! A central core is made up of the living, dining and family rooms, plus the kitchen. Each functions with an outdoor living area. The younger generation has its sleeping zone divorced from the master bedroom. The location of the attractive attached garage provides direct access to the front entry. Don't miss the vanity, the utility room with laundry equipment, the snack bar and the raised hearth fireplace. Note three pass-thrus from the kitchen. Observe the beamed and sloping ceilings of the living areas.

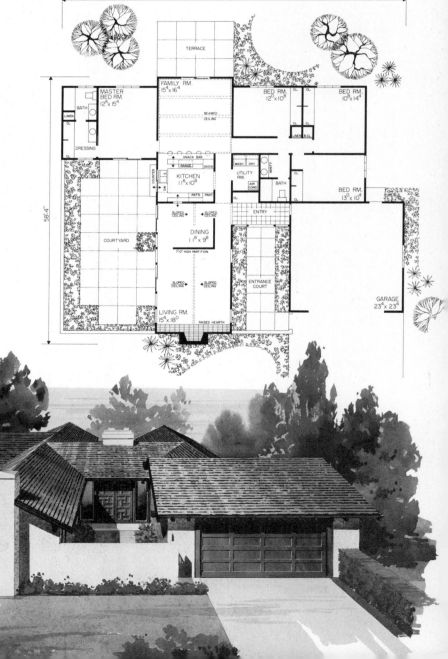

Design 32743
1,892 Sq. Ft.; 23,300 Cu. Ft.

● For those who feel they really don't require both a living and a family room, this refreshing contemporary will serve its occupants well, indeed. Ponder deeply its space and livability; for this design makes a lot of economic sense, too. First of all, placing the attached garage at the front cuts down on the size of a site required. It also represents an appealing design factor. The privacy wall and overhead trellis provide a pleasant front courtyard. Inside, the gathering room satisfies the family's more gregarious instincts, while there is always the study nearby to serve as a more peaceful haven. The separate dining room and the nook offer dining flexibility. The two full baths highlight the economical back-to-back plumbing feature. Note the rear terraces.

Design 32330
1,854 Sq. Ft.; 30,001 Cu. Ft.

● Your family will never tire of the living patterns offered by this appealing home with its low-pitched, wide overhanging roof. The masonry masses of the exterior are pleasing. While the blueprints call for the use of stone, you may wish to substitute brick veneer. Sloping ceiling and plenty of glass will assure the living area of a fine feeling of spaciousness. The covered porches enhance the enjoyment of outdoor living. Two baths serve the three bedroom sleeping area.

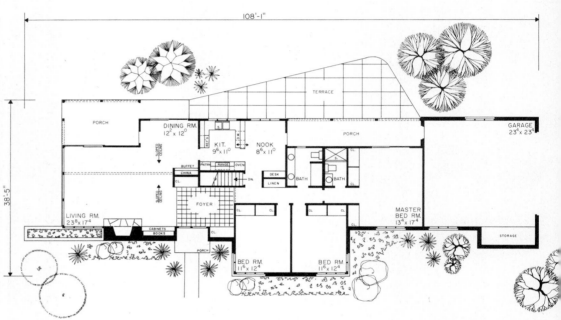

Design 33165
1,940 Sq. Ft.; 20,424 Cu. Ft.

● A practical plan with a contemporary facade. The center entrance encourages fine traffic circulation. To the right of the entry hall is the formal living area with its sunken living room and separate dining room separated by an attractive built-in planter. To the left of the entry hall is the three bedroom, two full bath sleeping zone. At the end of the entrance hall is the informal living area featuring the family room, efficient kitchen and laundry room.

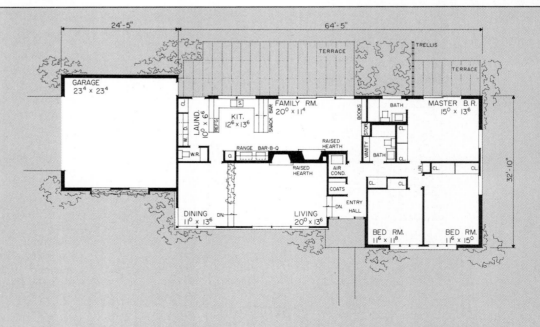

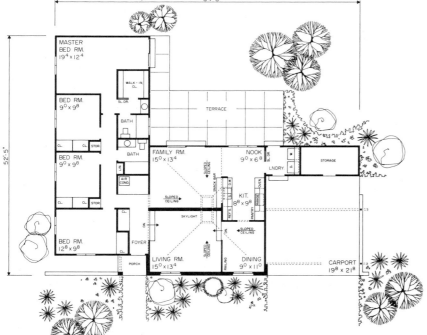

Design 32312
1,703 Sq. Ft.; 18,801 Cu. Ft.

● If you like contemporary living patterns to go with your refreshingly distinctive exterior, this four bedroom design may be the perfect choice for you and your family. Essentially of frame construction with vertical siding, brick veneer is used prudently to foster an appealing contrast in exterior materials. The covered front entrance leads to the centered foyer which effectively routes traffic to the well-zoned interior. The living and dining rooms function together for formal entertaining. The family room, nook and kitchen also function well together. This informal living zone is readily accessible to the rear terrace.

Design 31917
1,728 Sq. Ft.; 32,486 Cu. Ft.

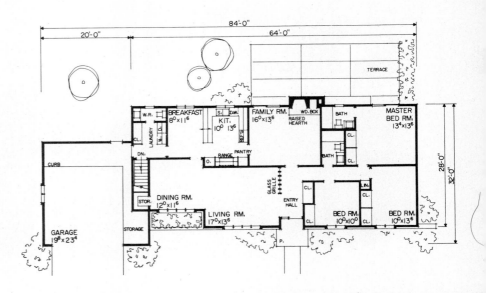

● Imagine your family living in this appealing one-story home. Think of how your living habits will adjust to the delightful patterns offered here. Flexibility is the byword; for there are two living areas — the front, formal living room and the rear, informal family room. There are two dining areas — the dining room overlooking the front yard and the breakfast room looking out upon the rear yard. There are outstanding bath facilities — a full bath for the master bedroom and one that will be handy to the living areas as a powder room. Then there is the extra wash room just where you need it — handy to the kitchen, the basement and the outdoors.

Design 31891
1,986 Sq. Ft.; 23,022 Cu. Ft.

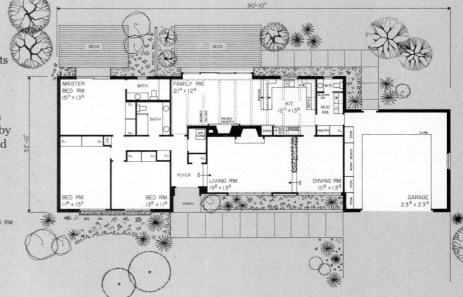

● There is much more to a house than just its exterior. And while the appeal of this home would be difficult to beat, it is the living potential of the interior that gives this design such a high ranking. The sunken living room with its adjacent dining room is highlighted by the attractive fireplace, the raised planter and the distinctive glass panels. A raised hearth fireplace, snack bar and sliding glass doors which open to the outdoor deck are features of the family room. The work center area is efficient. It has plenty of storage space and a laundry area.

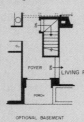

Design 31396
1,664 Sq. Ft.; 30,229 Cu. Ft.

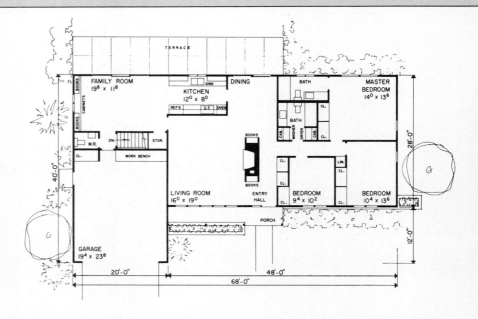

● Three bedrooms, 2½ baths, a formal dining area, a fine family room and an attached two-car garage are among the highlights of this frame home. The living-dining area is delightfully spacious with the fireplace wall, having book shelves at each end, functioning as a practical area divider. The many storage units found in this home will be a topic of conversion. The cabinets above the strategically located washer and dryer, the family room storage wall and walk-in closet and the garage facilities are particularly noteworthy. Blueprints show how to build this house with and without a basement.

Design 31237
1,616 Sq. Ft.; 15,150 Cu. Ft.

● The distinctiveness of contemporary design takes many forms. Here the trim flat roofs set the character aided by the vertical siding, the wood posts and the masses of glass. The front entranceway is covered and is enhanced by the picturesque court. The plan is equally unique since it features the two children's bedrooms functioning with the family room and the parents' master bedroom functioning with the formal living room. Functional terraces play their part: the play terrace is adjacent to the children's area; the quiet terrace is next to the parents' room and the living terrace services the dining and living rooms.

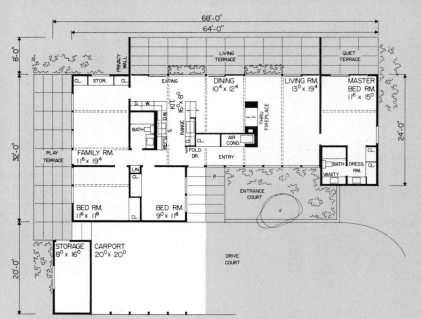

Design 32355
1,892 Sq. Ft.; 19,300 Cu. Ft.

● Whether you build this contemporary out in the country or on a relatively narrow suburban site, it will matter very little. And it will not diminish the unique livability a bit! Here in less than 1,900 square feet are four bedrooms, two living areas, a separate dining room and an outstanding kitchen. The major rooms, excluding the kitchen, have sloping ceilings. Sliding glass doors permit the bedrooms and the living and family rooms to function with the outdoor pool and terrace. Notice the raised hearth fireplace, the snack bar and the walk-in closet of the master bedroom. The utility room houses the laundry equipment.

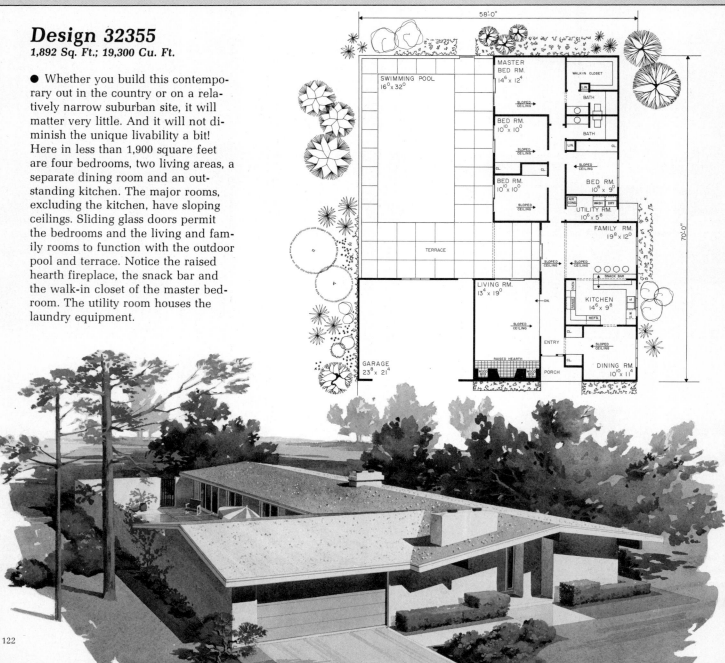

Design 32234
1,857 Sq. Ft.; 22,748 Cu. Ft.

● Planned to assure privacy from the street, this contemporary home sacrifices nothing in the way of natural light. Plenty of sliding glass doors and plastic skylights provide an abundance of daytime light. The zoning is interesting and practical. The center foyer is backed-up by the utility area and the kitchen. To one side is the sleeping zone; to the other, the living zone. The master bedroom features a huge walk-in closet and private bath. A handy corner of the family room will provide the space for family dining. The varied activities of the family can be pursued with plenty of space left over. Note storage wall. Two-way fireplace separates family and living rooms.

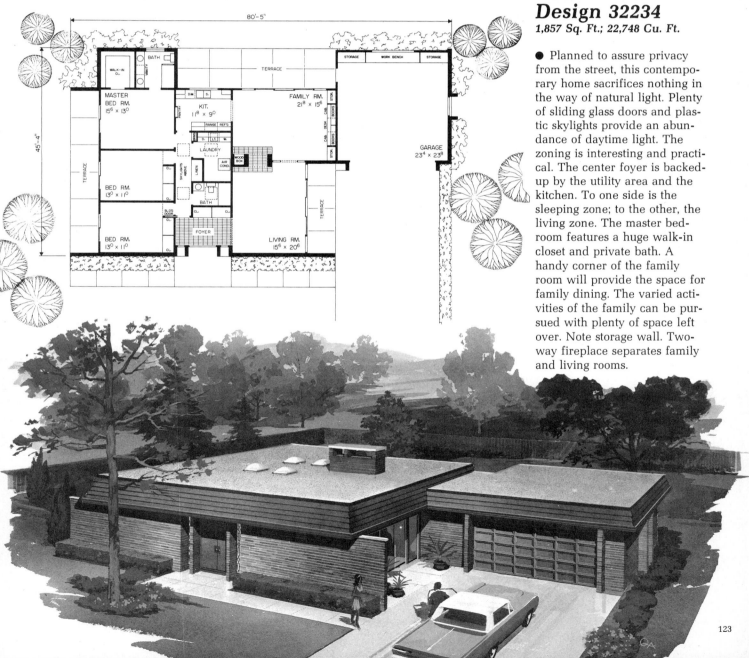

Design 31869
1,776 Sq. Ft.; 27,627 Cu. Ft.

● What will you like best about this design? It may be difficult for many to decide upon their favorite features. High on everyone's list of features will surely be: The refreshingly simple contemporary exterior and the tremendous amount of livability within 1,776 square feet. Among the specific highlights are the low-pitched roof, the pleasant window treatment, the raised planters, the spacious front living room, the separate dining room and the family room with fireplace and built-in book shelves.

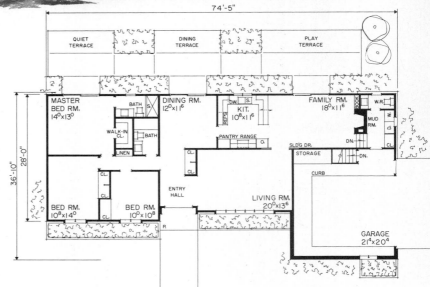

Design 31760
1,680 Sq. Ft.; 17,304 Cu. Ft.

● Of modest size, this L-shaped ranch home will be built most economically. The perfectly rectangular shape of the living portion features an exceptional amount of livability. The large family will find its return per construction dollar a tremendously good investment. Why? Consider these features: Three fine bedrooms for the children; a master bedroom with bath; a family room functioning ideally with the efficient kitchen and the rear terrace; a formal dining room and a spacious living area.

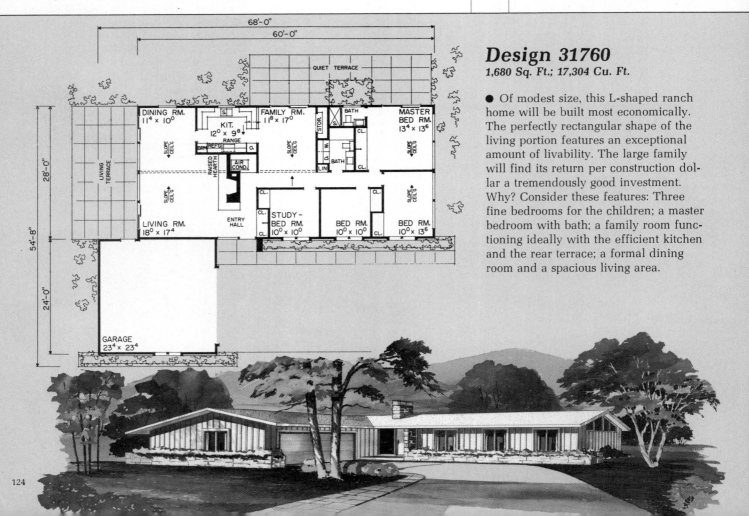

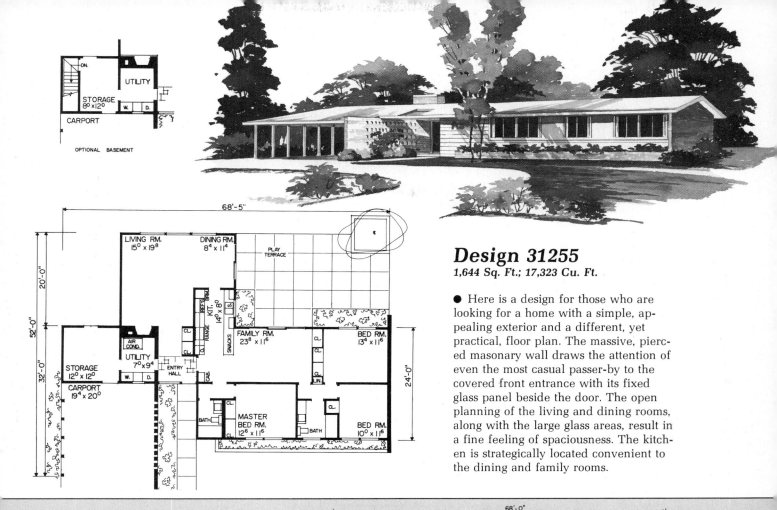

Design 31255
1,644 Sq. Ft.; 17,323 Cu. Ft.

● Here is a design for those who are looking for a home with a simple, appealing exterior and a different, yet practical, floor plan. The massive, pierced masonary wall draws the attention of even the most casual passer-by to the covered front entrance with its fixed glass panel beside the door. The open planning of the living and dining rooms, along with the large glass areas, result in a fine feeling of spaciousness. The kitchen is strategically located convenient to the dining and family rooms.

Design 31759
1,618 Sq. Ft.; 16,455 Cu. Ft.

● This economically built contemporary home boasts efficiently planned living space and handsome design. The low-pitched, wide over-hanging roof, the raised planter, the double front doors and the simplicity of the glass areas, all contribute to making this a home to be proud of. For the large family, there are four bedrooms and two baths in the sleeping area. The center entry hall features a clothes closet and a miscellaneous storage closet. Another storage unit projects to the terrace area.

Design 32380 1,997 Sq. Ft.; 18,784 Cu. Ft.

● This flat roofed contemporary with its impressive masses of brick fosters an aura of quiet simplicity. From the street it enjoys the utmost in privacy. The projection of the carport to the front adds interest to the exterior. In addition, it permits the use of a much smaller parcel of land. The zoning in this home is, indeed, interesting. The four bedroom, two bath sleeping zone will enjoy its privacy. The formal living room will experience no unnecessary traffic. The family room will be the hub of activities. It functions with the outdoor terrace and has a commanding raised hearth fireplace. The kitchen is nearby and is flanked by the formal and informal eating areas.

Design 32349
1,863 Sq. Ft.; 17,505 Cu. Ft.

● This plan can be built with two-strikingly contemporary facades. The living patterns of this refreshingly simple design will be delightfully different. From the formal sunken front living room to the informal beamed ceilinged rear family room, this interior has much to offer. There is a formal dining room with a convenient pass-thru from the efficient kitchen. Then there is the snack bar, also but an arms reach from the kitchen. Back-to-back plumbing is an economical feature of the two full baths. The focal point of the family room is the raised hearth fireplace flanked by bookshelves, cabinet and wood box. Sliding glass doors provide easy access to the outdoor patio from the master bedroom and family room. The blueprints ordered for this design show details for both exteriors as well as for basement and non-basement construction.

OPTIONAL BASEMENT

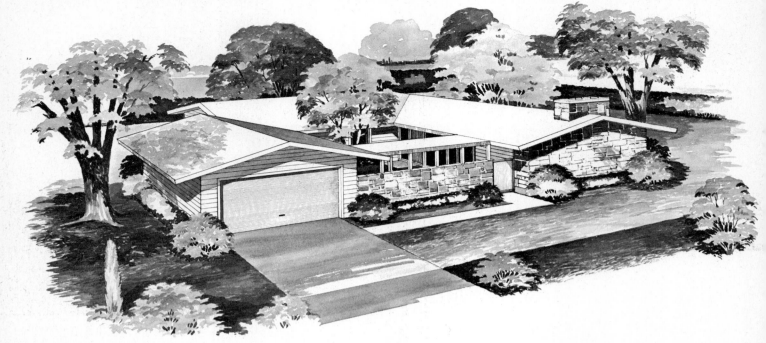

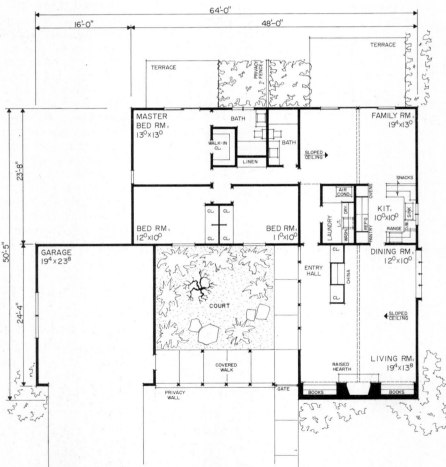

Design 31932 1,678 Sq. Ft.; 17,115 Cu. Ft.

● Here is a unique contemporary design whose projecting wings reach out to enclose a big court. The high privacy wall helps support the covering for the walkway inside. This enchanting area will be enjoyed the year' round. The windows in the living room permit a fine view of the court area and the undistrubed beauty of its landscaping whatever the season. The rear terraces off the master bedroom and the family room are separated by planting areas and a privacy fence. A study of the basic L-shaped floor plan reveals convenient living patterns. The informal family room and the formal living/dining area feature sloping ceilings for a feeling of spaciousness.

Design 31947 1,764 Sq. Ft.; 18,381 Cu. Ft.

● When it comes to housing your family, if you are among the contemporary-minded, you'll want to give this L-shaped design a second, then even a third, or fourth, look. It is available as either a three or four bedroom home. If you desire the three bedroom, 58 foot wide design order blueprints for 31947; for the four bedroom, 62 foot wide design, order 31948. Inside, you will note a continuation of the contemporary theme with sloping ceilings, exposed beams and a practical 42 inch high storage divider between the living and dining rooms. Don't miss the mud rooms.

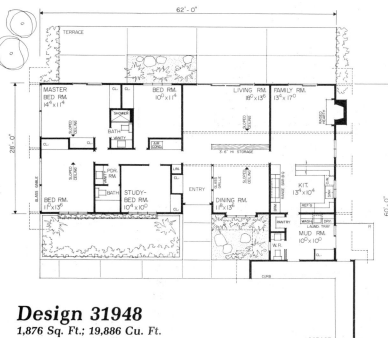

Design 31948
1,876 Sq. Ft.; 19,886 Cu. Ft.

Design 32797
1,791 Sq. Ft.; 37,805 Cu. Ft.

● The exterior appeal of this delightful one-story contemporary is sure to catch the attention of all who pass by. The hipped roof adds an extra measure of shading along with the privacy wall which shelters the front court and entry. The floor plan also will be outstanding to include both leisure and formal activities. The gathering room has a sloped ceiling, sliding glass door to rear terrace and a thru-fireplace to the family room. This room also has access to the terrace and it includes the informal eating area. A pass-thru from the U-shaped kitchen to the eating area makes serving a breeze. Formal dining can be done in the front dining room. The laundry area is adjacent to the kitchen and garage and houses a wash room. Peace and quiet can be achieved in the study. The sleeping zone consists of three bedrooms and two full back-to-back baths. Additional space will be found in the basement.

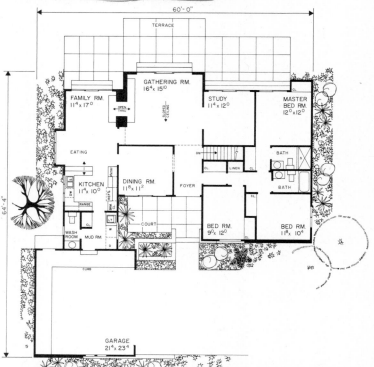

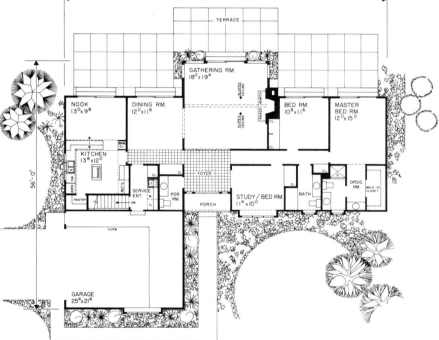

Design 32792
1,944 Sq. Ft.; 37,505 Cu. Ft.

● Indoor-outdoor living could hardly be improved upon in this contemporary design. All of the rear rooms have sliding glass doors to the large terrace. Divide the terrace in three parts and the nook and dining room have access to a dining terrace, the gathering room to a living terrace and two bedrooms to a lounging terrace. A delightful way to bring the outdoors view inside. Other fine features include the efficient kitchen which has plenty of storage space and an island range, a first floor laundry with stairs to the basement and a powder room adjacent to the front door.

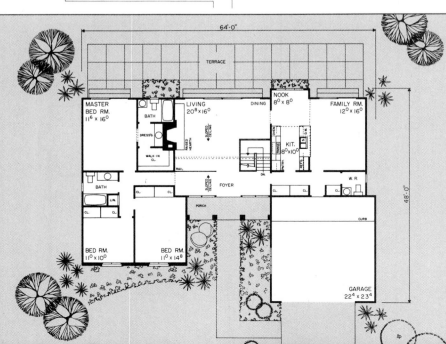

Design 32528
1,754 Sq. Ft.; 37,832 Cu. Ft.

● This inviting U-shaped western ranch adaptation offers outstanding living potential behind its double front doors and flanking glass panels. In but 1,754 square feet there are three bedrooms, 2½ baths, a formal living room and an informal family room, an excellently functioning interior kitchen, an adjacent breakfast nook and good storage facilities. The open stairwell to the lower level basement can be an interesting, interior feature. Note raised hearth fireplace and sloped ceiling.

Design 31798
1,720 Sq. Ft.; 17,280 Cu. Ft.

● Has your family ever wished they lived in a house where there could be a complete separation of the parents' and the children's living and sleeping facilities? If so, consider this L-shaped house. You will note here that such a separation does not neccessitate any huge or palatial-size floor plan. Clever planning has resulted in this fine, modest-size plan. Try to visualize how your family will function in this house. The two full baths service their respective areas well. The family room will be the teenagers' informal activities area. Sloping ceilings enhance the spaciousness of the interior. Free-standing built-in cabinet work is open above 7½ feet and carries over the doorways in the form of a valance. This is just one of the many features which will attract you to this fine design.

Design 31785 1,868 Sq. Ft.; 22,589 Cu. Ft.

● A refreshing exterior which will be like a breath of fresh air to your new neighborhood. This exquisite facade is reminiscent of Japanese architecture. Among the exterior features is the *Irimoya* roof, a combination of gable and hip styles. The wide eaves, *sori*, protect against the rain and the high, hot sun of summer, yet let in the low winter sunlight. The lattice-work grilles at the front resemble the *shoji* screens that the Japanese use on the exteriors of their homes. A study of the floor plan reveals a fine awareness of the outdoors. The interior features both formal and informal living areas. Kitchen eating space, plus a formal dining room is provided. There are plenty of bath and storage facilities.

Design 31019
1,996 Sq. Ft.; 20,550 Cu. Ft.

● Here is an attention-getter wherever you build it—in town, or on the outskirts of town. The curving drive approach has a way of establishing an impressive setting. The roofs are pitched low and they overhang to create that low-slung effect. The window treatment is simple, yet appealing, while the natural stone masses and the planting enhance the visual impact. The double front doors are recessed to provide shelter during inclement weather. The kitchen-family room area looks out upon its own side terrace, while the formal living room and bedrooms have their rear terrace.

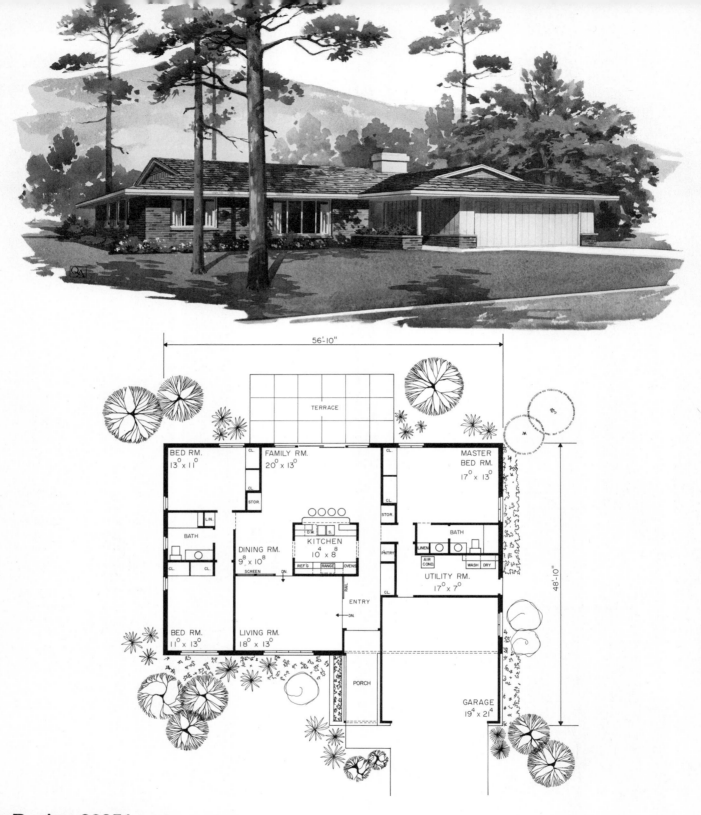

Design 32351 1,862 Sq. Ft.; 22,200 Cu. Ft.

● The extension of the wide overhanging roof of this distinctive home provides shelter for the walkway to the front door. A raised brick planter adds appeal to the outstanding exterior design. The living patterns offered by this plan are delightfully different, yet extremely practical. Notice the separation of the master bedroom from the other two bedrooms. While assuring an extra measure of quiet privacy for the parents, this master bedroom location may be ideal for a live-in-relative. Locating the kitchen in the middle of the plan frees up valuable outside wall space and leads to interesting planning. The front living room is sunken for dramatic appeal and need not have any cross-room traffic. The utility room houses the laundry and the heating and cooling equipment.

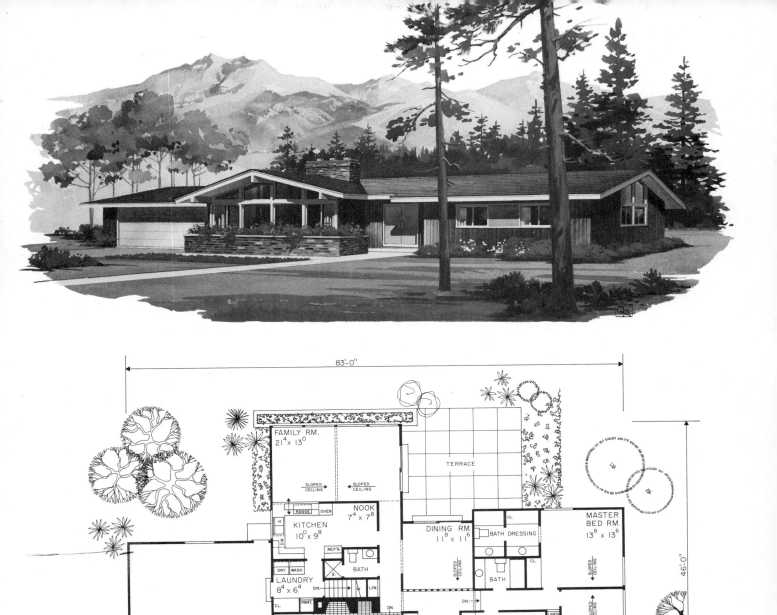

Design 32363 1,978 Sq. Ft.; 27,150 Cu. Ft.

● You will have a lot of fun deciding what you like best about this home with its eye-catching glass-gabled living room and wrap-around raised planter. A covered porch shelters the double front doors. Projecting to the rear is a family room identical in size with the formal living room. Between these two rooms there are features galore. There is the efficient kitchen with pass-thru and informal eating space. Then, there is the laundry with a closet, pantry and the basement stairs nearby. Also, a full bath featuring a stall shower. The dining room has a sloped ceiling and an appealing, open vertical divider which acts as screening from the entry. The three bedroom, two bath sleeping zone is sunken. The raised hearth fireplace in the living room has an adjacent wood box.

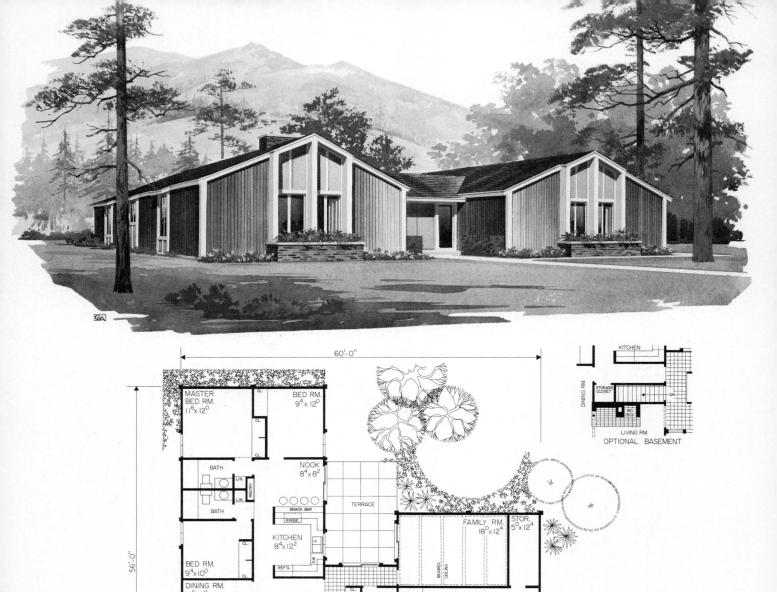

Design 32382 1,755 Sq. Ft.; 20,114 Cu. Ft.

● Here is a refreshing, modified H-shaped home with a most interesting interior. The spacious formal living and dining area has a sloped ceiling and a commanding raised hearth fireplace. Back-to-back plumbing highlights the two full baths. The kitchen and nook look out upon, and function with, the outdoor terrace. The entry unit is the connecting link to the all-purpose family room. It is spacious and provides direct access to the garage as well as, through sliding glass doors, to the partially enclosed terrace. Note the wardrobe closet and the adjacent powder room. The family room with its beamed ceiling functions with the private terrace. A bulk storage room off the garage is just the place for all that lawn and gardening paraphernalia. Observe optional basement.

Design 32332 1,992 Sq. Ft.; 32,674 Cu. Ft.

● This contemporary one-story design really functions on two levels. The foyer, featuring high sloping ceiling and built-in planting area, routes traffic on the same level to the three bedroom, two bath sleeping zone. Then up four steps are the living areas. And what fine facilities await! Separating the living and family rooms is the thru fireplace which may be enjoyed equally from either room. Large glass areas and sliding glass doors permit optimum awareness of the rear outdoor terrace and its surroundings. Observe the built-in cabinets, book shelves, desk and wood box. The kitchen is only a step from the dining room and has a pass-thru to the snack bar. Note laundry area which houses a wash room, basement stairs, pantry, closet and access to garage.

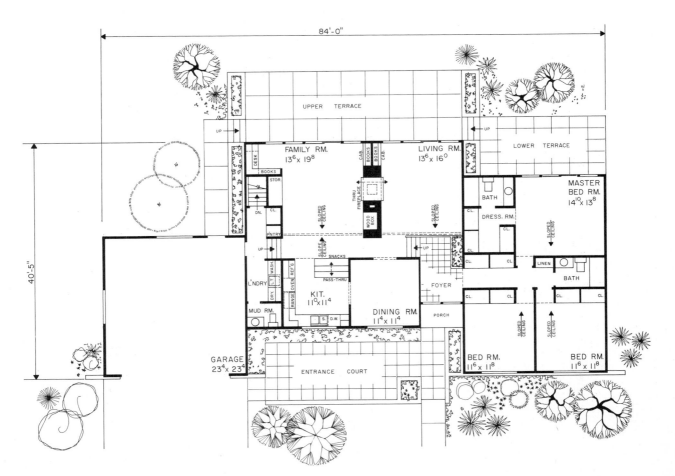

Design 32795

1,952 Sq. Ft.; 43,500 Cu. Ft.

● This three-bedroom design leaves no room for improvement. Any size family will find it difficult to surpass the fine qualities that this home offers. Begin with the exterior. A fine contemporary design with open trellis work above the front covered private court, this area is sheltered by a privacy wall extending from the projecting garage. Inside, the floor plan will be just as breathtaking. Begin at the foyer and choose a direction. To the right is the sleeping wing equipped with three bedrooms and two baths. Straight ahead from the foyer is the gathering room with thru-fireplace to the dining room. To the right is the work center. This area includes a breakfast nook, a U-shaped kitchen and laundry.

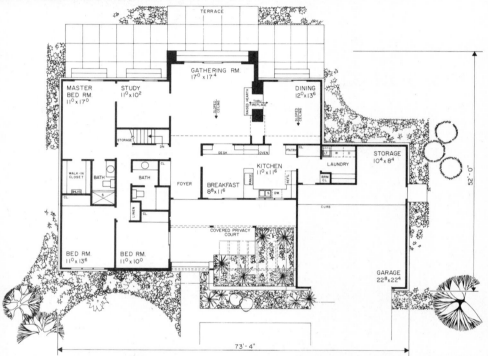

Design 32754
1,844 Sq. Ft.; 26,615 Cu. Ft.

● This really is a most dramatic and refreshing contemporary home. The slope of its wide overhanging roofs is carried right indoors to provide an extra measure of spaciousness. The U-shaped privacy wall of the front entrance area provides an appealing outdoor living spot accessible from the front bedroom. The rectangular floor plan will be economical to build. Notice the efficient use of space and how it all makes its contribution to outstanding livability. The small family will find its living patterns delightful, indeed. Two bedrooms and two full baths comprise the sleeping zone. The open planning of the L-shaped living and dining rooms is most desirable. The thru-fireplace is just a great room divider. The kitchen and breakfast nook function well together. There is laundry and mechanical room nearby.

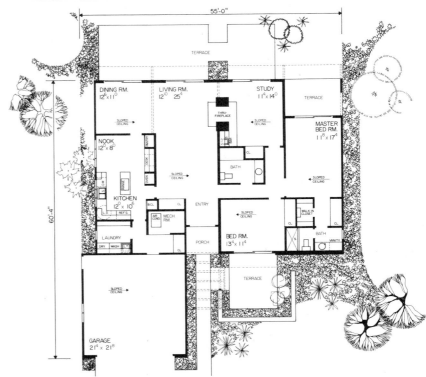

Design 32796
1,828 Sq. Ft.; 39,990 Cu. Ft.

● This home features a front living room with sloped ceiling and sliding glass doors which lead to a front private court. What a delightful way to introduce this design. This bi-nuclear design has a great deal to offer. First - the children's and parent's sleeping quarters are on opposite ends of this house to assure the utmost in privacy. Each area has its own full bath. The interior kitchen is a great idea. It frees up valuable wall space for the living areas exclusive use. There is a snack bar in the kitchen/family room for those very informal meals. Also, a planning desk is in the family room. The dining room is conveniently located near the kitchen plus it has a built-in china cabinet. The laundry area has plenty of storage closets plus the stairs to the basement. This home will surely be a welcome addition to any setting.

Design 31215
1,696 Sq. Ft.; 16,672 Cu. Ft.

● Whatever the view of this refreshing contemporary home, it will surely be a dramatic one. The low-pitched overhanging roof, the exposed rafter tails, the glass gabled ends and the sliding glass doors are all delightful features. Note that all areas function with the outdoors. There are four terraces to serve the family who enjoys the outdoors. A living terrace adjacent to the living/dining area, a play terrace off the family room, a quiet terrace for the master bedroom and yet another accessible from the other two bedrooms. The play terrace has a barbecue unit built in.

Design 31813
1,755 Sq. Ft.; 17,638 Cu. Ft.

● A modest sized home with a delightfully refreshing contemporary exterior and a functional family floor plan. The low roofs with the wide overhangs, the dramatic glass areas and the appealing privacy wall, which helps form a spacious entry court, are among the desirable features of the exterior design. The livability of this floor plan is sure to fit your family's needs. This is a "bi-nuclear" design. Meaning that the master bedroom has been completely separated from the remaining bedrooms. In this case, the master bedroom is in the formal living area and the family bedrooms near the informal family room.

Design 31390
1,664 Sq. Ft.; 17,338 Cu. Ft.

● What will it be like touring the inside of this home? It will, indeed, be fun. Worthy of immediate comment are the traffic patterns emanating from the entry hall. The sleeping wing which projects toward the street has four bedrooms and two full baths.

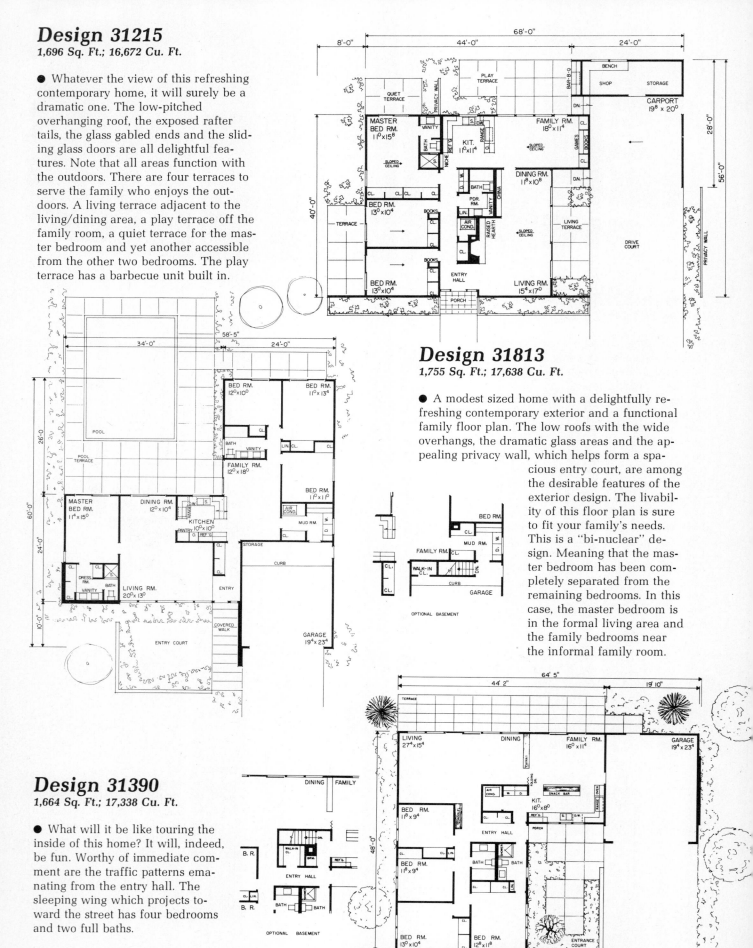

Design 31126
1,803 Sq. ft.; 18,390 Cu. Ft.

● Good living and good design go together. Here, a gently sloping, wide overhanging roof covers a unique and practical floor plan. Excellent glass areas provide for an abundance of natural light to an already spacious interior with sloping ceilings. Sliding glass doors lead from the bedroom area, the formal living area and the informal family room area to the outdoor terraces. A dramatic fireplace with raised hearth is the focal point of the fine living room.

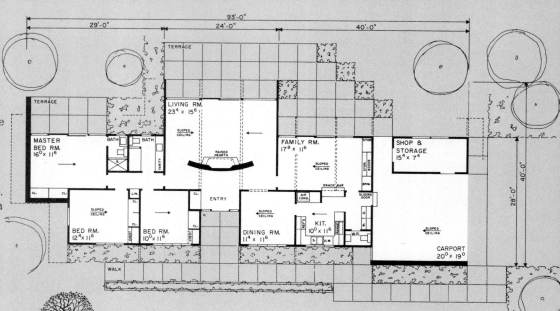

Design 33181
1,920 Sq. Ft.; 29,515 Cu. Ft.

● If yours is good taste with a flair for the unique, then your choice of this design will result in years of satisfaction. This dramatic home has a facade all its own and an interior to match. The covered front porch shelters double front doors which open to an interior in which it is going to be fun to live. Sloping ceilings heighten the spaciousness of the living areas. An exciting thru-fireplace may be enjoyed from both the living and dining rooms. The family room with its snack bar will be in constant use. Sliding glass doors lead to the living terrace and the dining porch. The mud room and wash room are placed in the most strategic areas. Other features include two full baths, a built-in planter and a partial basement.

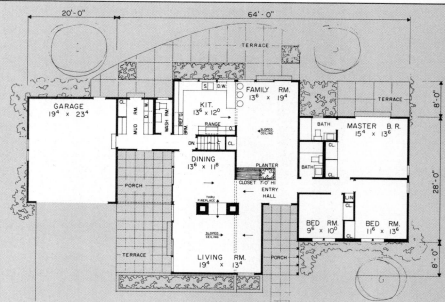

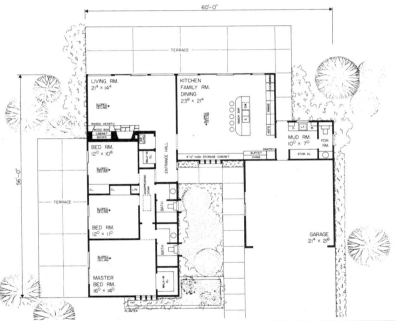

Design 32238
1,900 Sq. Ft.; 21,945 Cu. Ft.

● Design distinction comes in many forms as this unique contemporary will attest. Horizontal siding, shed-type roof, accented vertical window lines and an irregular shape are the features which help create such a refreshing image. The raised stone planter adds that extra measure of appeal. Inside, there is a truly different type of floor plan. From the entrance hall traffic can flow directly to the three bedroom, two bath sleeping area; the quiet living room; the kitchen/family room/dining area. All areas have sliding glass doors and function directly with an outdoor terrace. Sloped ceilings add an extra dimension of spaciousness. As an extension of the garage, the mud room and powder room could hardly be more strategically located.

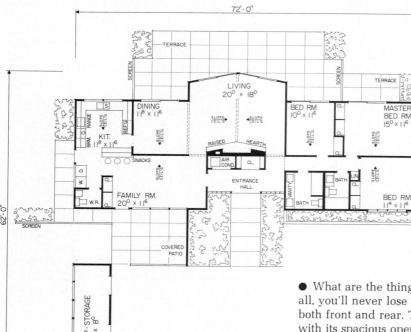

Design 31096
1,787 Sq. Ft.; 17,262 Cu. Ft.

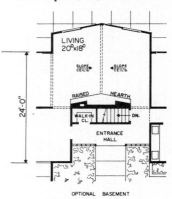

● What are the things you'll like best about this design? Well, first of all, you'll never lose your enthusiasm for its contemporary exterior—both front and rear. Then, the cheerful environment of the interior with its spacious open planning, sloping beamed ceilings and large glass areas will provide you with a constant sense of well-being. List other appealing features. The blueprints you order for this design show details for basement and non-basement. Note basement stair location.

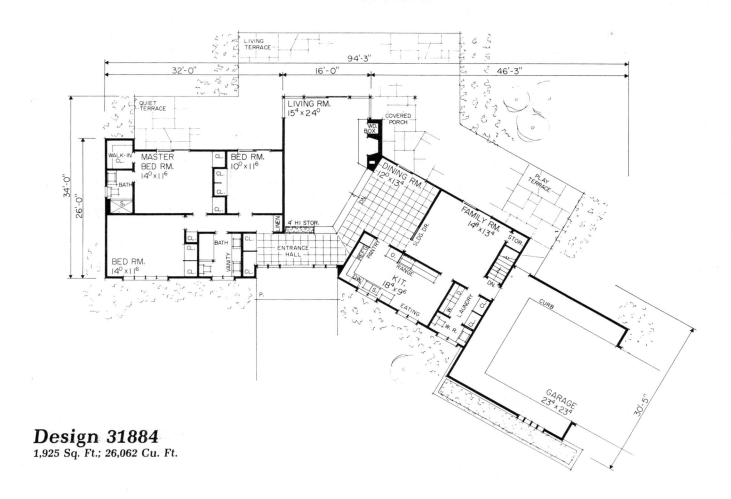

Design 31884
1,925 Sq. Ft.; 26,062 Cu. Ft.

● If you are searching for something with an air of distinction both inside and out then search no more. You could hardly improve upon what this home has to offer. You will forever be proud of the impressive hip-roof, angular facade. Its interest will give it an identity all its own. As for the interior, your everyday living patterns will be a delight, indeed. And little wonder, clever zoning and a fine feeling of spaciousness set the stage. As you stand in the entrance hall, you look across the tip of a four foot high planter into the sunken living room. Having an expanse of floor space, the wall of windows and the raised hearth fireplace, the view will be dramatic. Notice covered porch, play terrace and quiet terrace which will provide great outdoor enjoyment.

Design 33139
1,680 Sq. Ft.; 17,808 Cu. Ft.

● Whatever the setting - in the mountains or on the flat lands - this contemporary will be an attractive addition. While this house guards its privacy from the street, it promotes wonderful indoor-outdoor relationship to the rear. A study of the plan reveals five sliding glass doors. In addition to the functional rear terrace areas, there is the court and side terrace each of which will enjoy seclusion resulting from the patterned masonry walls. The interior is extremely interesting and practical. Notice the separation of the children's and parents' bedrooms. A full bath serves each. The centrally located U-shaped kitchen is adjacent to the breakfast eating area, the snack bar and the family dining room.

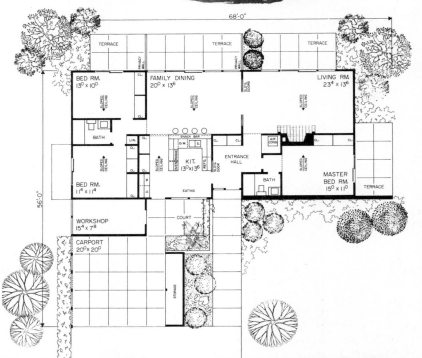

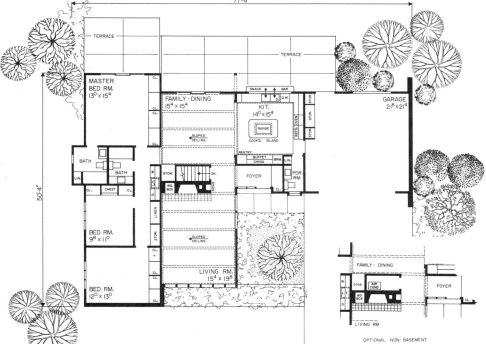

Design 32201
1,811 Sq. Ft.; 32,708 Cu. Ft.

● This appealing L-shaped contemporary with its interesting glass areas, raised planter and prudent use of contrasting exterior materials has much to offer the modest sized family. Its fine zoning will lead to pleasurable living patterns. Notice the sleeping wing. There are two sizable bedrooms, plus a smaller bedroom which may instead be a study, sewing room or a guest room. There are two baths and positively extraordinary storage facilities. The living zone has a big living room and a family dining room. Sloping beamed ceilings highlight these rooms. The homemaker's kitchen will be a joy in which to function. A pass-thru leads to the snack bar.

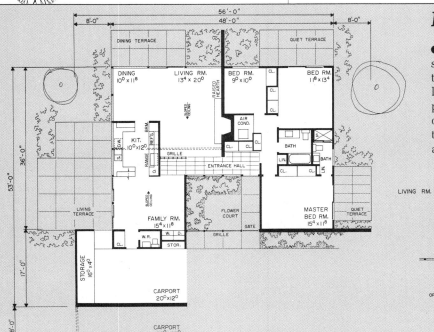

Design 31153 1,616 Sq. Ft.; 16,256 Cu. Ft.

● The simple appeal of this contemporary design is to be found in the solid brick masses, the delicate patterned grille work and the low-pitched roof. Inside the iron gate is the pleasing flower court. The projecting carport or garage, if you wish, offers the possibility of the development of an impressive drive court approach. Inside, the plan is certainly unique. The room relationships are excellent and the traffic patterns are orderly. You will love the indoor-outdoor accessibility. Blueprints for this particular design include details for both basement and non-basement construction. Don't miss the sloped ceilings, 2½ baths and all those closets.

147

Design 31129
1,904 Sq. Ft.; 34,066 Cu. Ft.

● Study this floor plan with care, for there is much to recommend it to the large family. The living room and separate dining area look out upon the rear yard. Sliding glass doors permit the dining area to function with the terrace — a delightful indoor-outdoor living feature. An appealing two-way fireplace may be enjoyed from both the living and dining areas. The efficient U-shaped kitchen overlooks the front yard and is but a few steps from the entry hall as well as the service entrance and garage.

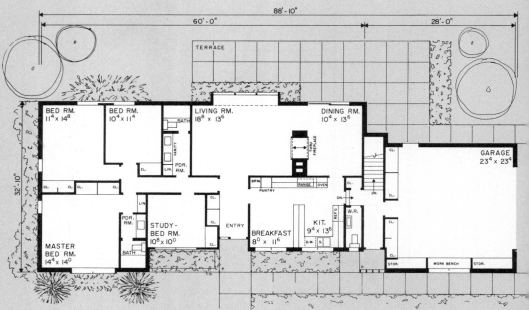

Design 33219
1,960 Sq. Ft.; 31,643 Cu. Ft.

● Here is a delightfully tailored contemporary with a low-pitched roof and a wide overhang. Its window treatment is simple and the brick masses are pleasing. The raised planters add that little extra touch thus making a pretty picture truly delightful. It would be difficult to predict which would be the favored spot — sitting before the fireplace in the living room, or the family room. Two full baths serve the three bedroom sleeping area, while an extra wash room is adjacent to the mud room.

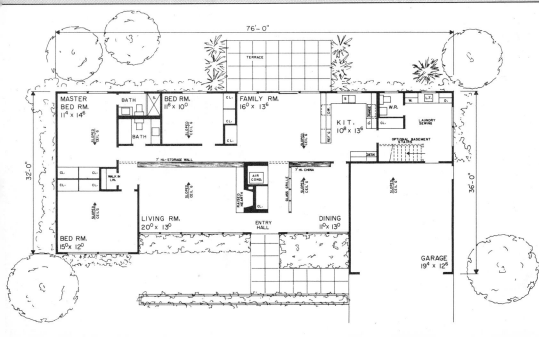

Design 31036
1,912 Sq. Ft.; 18,785 Cu. Ft.

● This pleasing contemporary design represents a fine combination of great exterior appeal and wonderful interior planning. The exterior is completely frame and features delightful glass areas protected by the wide overhanging roof. Although not a part of the house itself, the long planter adds notably to the overall appeal. The double doors of the center entry will be an attractive feature of the exterior. The flow of traffic throughout the plan will be very flexible and convenient.

Design 31867
1,692 Sq. Ft. — Excluding Atrium; 21,383 Cu. Ft.

● Looking for a new house involves a number of varied considerations. One that is most basic involves what you would like your family's living patterns to be. If you would like to introduce your family to something that is different and is sure to be fun for all, consider this dramatic atrium house. Parents and children alike will be thrilled by the experience and the enviroment engendered by this outdoor living area indoors. A skylight provides for the protection during inclement weather without restricting the flood of natural light. Notice how this atrium functions with the various areas. Observe relationships between children's rooms and family room; master bedroom and living room.

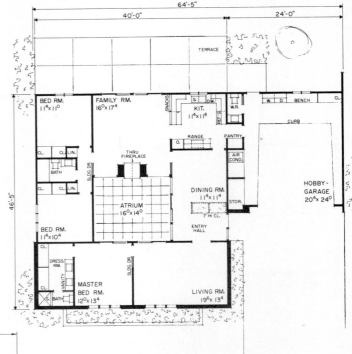

Design 32383
1,984 Sq. Ft. — Excluding Atrium; 20,470 Cu. Ft.

● Design your home around an atrium and you will enjoy living patterns unlike any you have experienced in the past. This interior area is assured complete outdoor privacy. Five sets of sliding glass doors enhance the accessibility of this unique area. With the two-car garage projecting from the front of the house, this design will not require a large piece of property. Worthy of particular note is the separation of the master bedroom from the other three bedrooms - a fine feature to assure peace and quiet. Side-by-side are the formal and informal living rooms. Both function with the rear terrace. Separating the two rooms is the thru-fireplace and double access wood box.

Design 31283
1,904 Sq. Ft. – Excluding Atrium; 18,659 Cu. Ft.

● Here is a unique home whose livable area is basically a perfect square. Completely adaptable to a narrow building site, the presence of the atrium permits the enjoyment of private outdoor living "indoors". Sliding glass doors open onto this delightful atrium with its attractive planting areas.

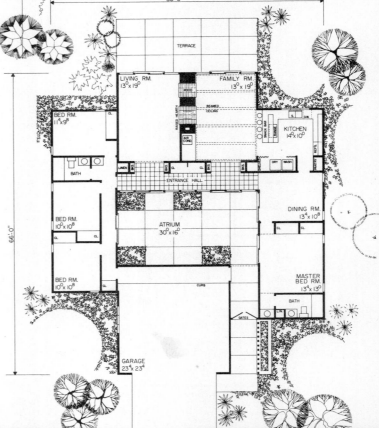

Design 33167
1,760 Sq. Ft.; 33,440 Cu. Ft.

● A modest sized home with all the attributes of a much larger home. The interior highlights are many indeed. There are three bedrooms and two full baths; an entry hall with two wardrobe storage areas; a 27 foot formal living and dining area with fireplace; a sizable family room with sliding glass doors to the rear terrace; an efficient kitchen with a built-in barbecue unit; a handy wash room; a full basement for bulk storage and an attached two-car garage. Note linen, broom and china storage facilities. A home to serve the family admirably for many a year to come.

Design 31765
1,924 Sq. Ft.; 22,092 Cu. Ft.

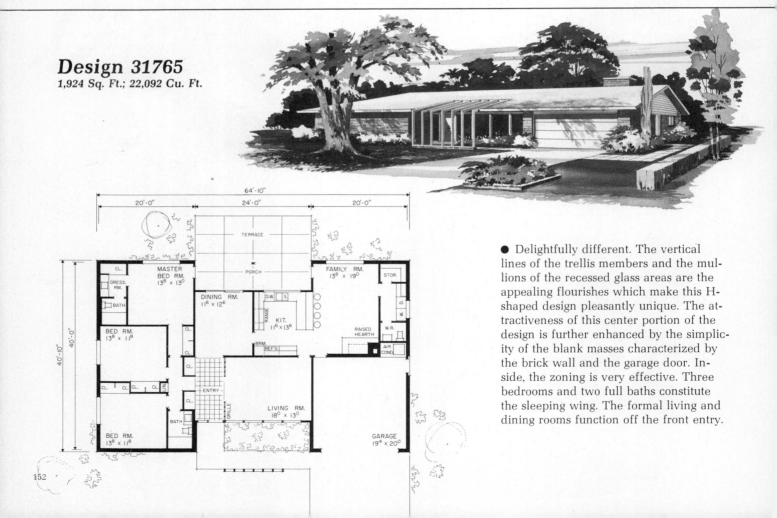

● Delightfully different. The vertical lines of the trellis members and the mullions of the recessed glass areas are the appealing flourishes which make this H-shaped design pleasantly unique. The attractiveness of this center portion of the design is further enhanced by the simplicity of the blank masses characterized by the brick wall and the garage door. Inside, the zoning is very effective. Three bedrooms and two full baths constitute the sleeping wing. The formal living and dining rooms function off the front entry.

Traditional & Contemporary
Homes for Restricted Budgets
Under 1200 Sq. Ft.

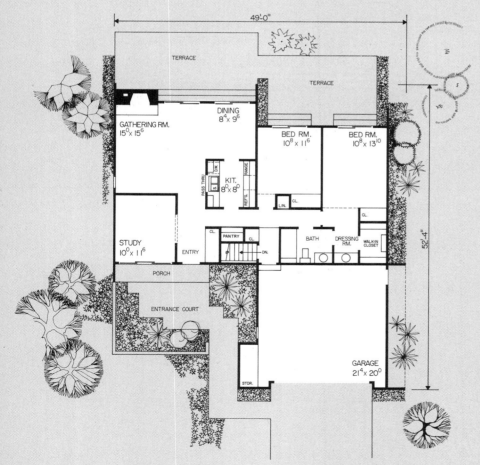

Design 32755
1,200 Sq. Ft.; 23,925 Cu. Ft.

● Here is truly an outstanding, low-cost design created to return all the pride of ownership and livability a small family or retired couple would ask of a new home. The living/dining area measures a spacious 23 feet. It has a fireplace and two sets of sliding glass doors leading to the large rear terrace. The two bedrooms also have access to this terrace. The kitchen is a real step-saver and has a pantry nearby. The study, which has sliding glass doors to the front porch, will function as that extra all-purpose room. Use it for sewing, guests, writing or reading or just plain napping. The basement offers the possibility for the development of additional recreation space. Note the storage area at the side of the garage. Many years of enjoyable living will surely be obtained in this home designed in the contemporary fashion.

Design 31297
1,034 Sq. Ft.; 11,324 Cu. Ft.

● The U-shape of this appealing home qualifies it for a narrow building site. The efficiency of the floor plan recommends it for convenient living during the retirement years. The charm of the exterior surely makes it a prize-winner. The attractive wood fence and its lamp post complete the enclosure of the flower court—a delightful setting for the walk to the covered front entrance. The work area is outstanding. There is the laundry with closet space, an extra wash room and a fine kitchen with eating space.

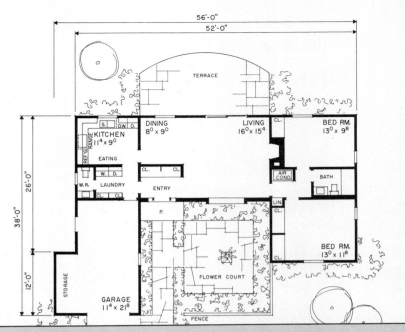

Design 31309
1,100 Sq. Ft.; 15,600 Cu. Ft.

● Here is a real low-cost charmer. Delightful proportion and an effective use of materials characterize this Colonial version. Vertical boards and battens, a touch of stone and pleasing window treatment catch the eye. The compact, economical plan offers spacious formal living and dining areas plus a family room. The kitchen is strategically located—it overlooks the rear yard and is but a few steps from the outdoor terrace. The attached garage has a large storage and utility area to the rear.

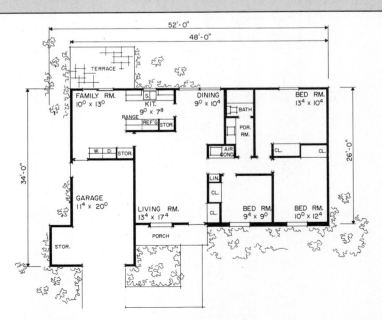

Design 31364
1,142 Sq. Ft.; 13,510 Cu. Ft.

● The family working within the confines of a restricted building budget will find this eye-catching traditional ranch home the solution to its housing needs. The brick exterior with its recessed front entrance, wood shutters, bowed window, attached garage and wood fence is charming, indeed. The living room is free of cross-room traffic and lends itself to effective and flexible furniture placement. The master bedroom has its own private bath with stall shower, while the main bath features a built-in vanity and adjacent linen storage.

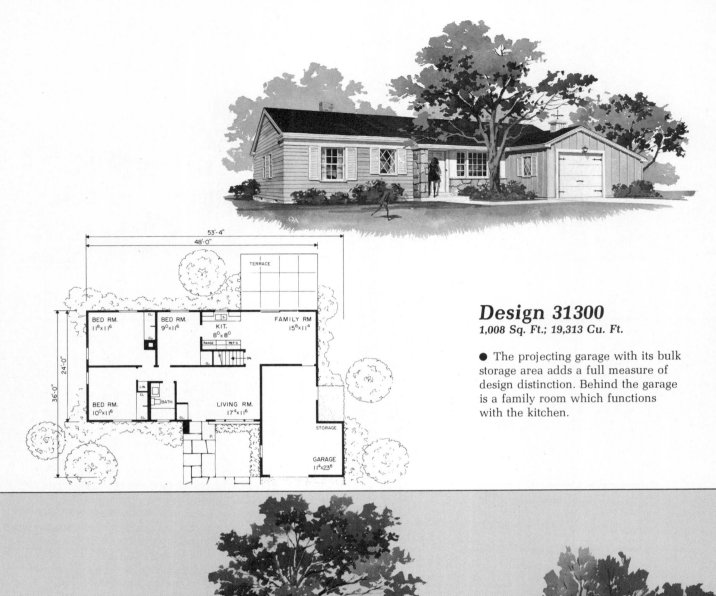

Design 31300
1,008 Sq. Ft.; 19,313 Cu. Ft.

● The projecting garage with its bulk storage area adds a full measure of design distinction. Behind the garage is a family room which functions with the kitchen.

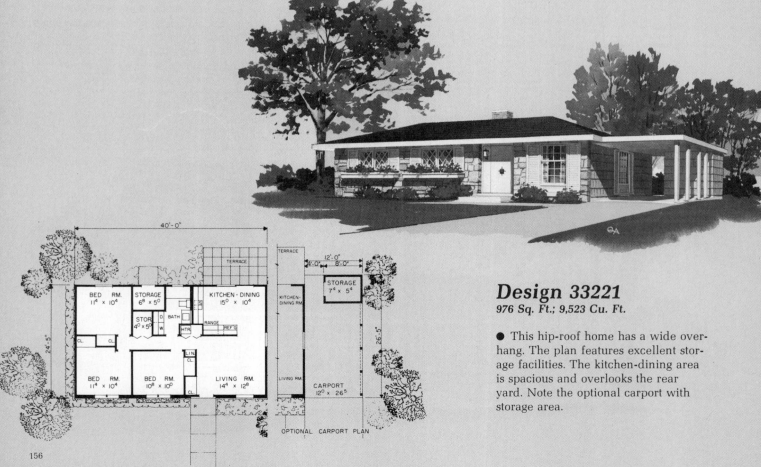

Design 33221
976 Sq. Ft.; 9,523 Cu. Ft.

● This hip-roof home has a wide overhang. The plan features excellent storage facilities. The kitchen-dining area is spacious and overlooks the rear yard. Note the optional carport with storage area.

Design 31301
1,056 Sq. Ft.; 19,536 Cu. Ft.

● Charming is just the word to describe this L-shaped traditional home. Note formal living and informal family rooms, U-shaped kitchen and extra wash room.

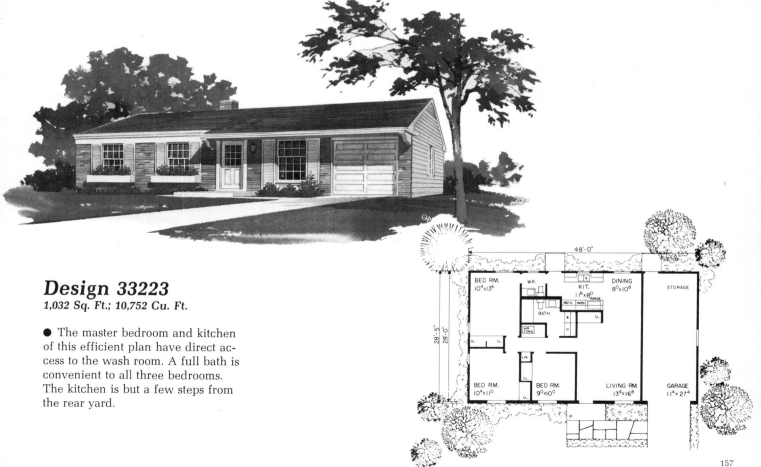

Design 33223
1,032 Sq. Ft.; 10,752 Cu. Ft.

● The master bedroom and kitchen of this efficient plan have direct access to the wash room. A full bath is convenient to all three bedrooms. The kitchen is but a few steps from the rear yard.

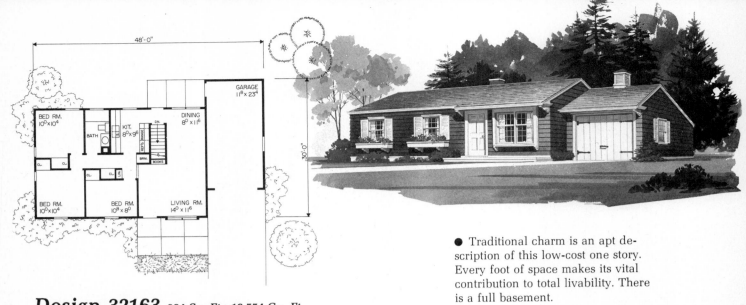

Design 32163 864 Sq. Ft.; 16,554 Cu. Ft.

● Traditional charm is an apt description of this low-cost one story. Every foot of space makes its vital contribution to total livability. There is a full basement.

● A hip-roofed contemporary with an attached carport. Both kitchen and dining room accessible to rear terrace. List the many storage units. Don't miss linen closets.

Design 32166 864 Sq. Ft.; 8,448 Cu. Ft.

● A compact, small home with its full measure of built-in efficiency. Inside bath frees valuable outside wall space for more effective planning. Three bedrooms.

Design 32164 864 Sq. Ft.; 9,245 Cu. Ft.

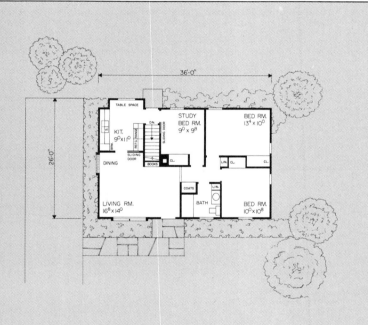

Design 32165
880 Sq. Ft.; 16,945 Cu. Ft.

● Whether called upon to function as a two or a three bedroom home, this attractive design will serve its occupants ideally for many years. There are two eating areas.

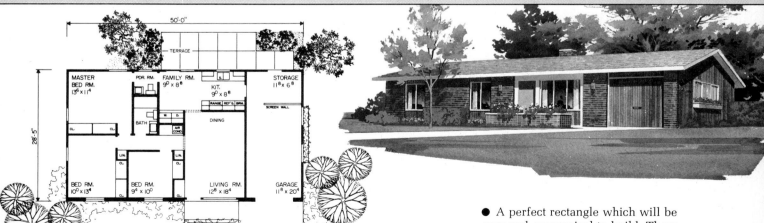

Design 32159 1,077 Sq. Ft.; 11,222 Cu. Ft.

● A perfect rectangle which will be easy and economical to build. There is an extra wash room, family kitchen, three bedrooms, full bath plus powder room and bulk storage area in the garage.

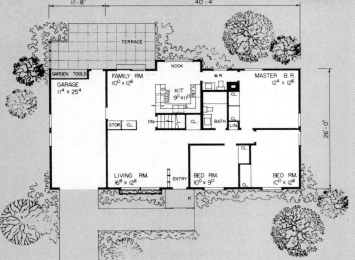

Design 33213
1,060 Sq. Ft.; 18,608 Cu. Ft.

● It would be difficult to find a design with more livability built into its 1,060 square feet. There are features galore. Make a list.

Design 31399
1,040 Sq. Ft.; 11,783 Cu. Ft.

● A real winner which will surely satisfy the restricted building budget, while returning a tremendous amount in the way of convenient living and pride of ownership.

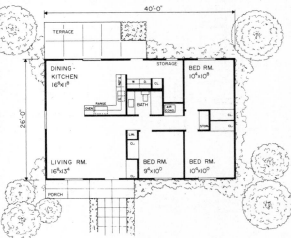

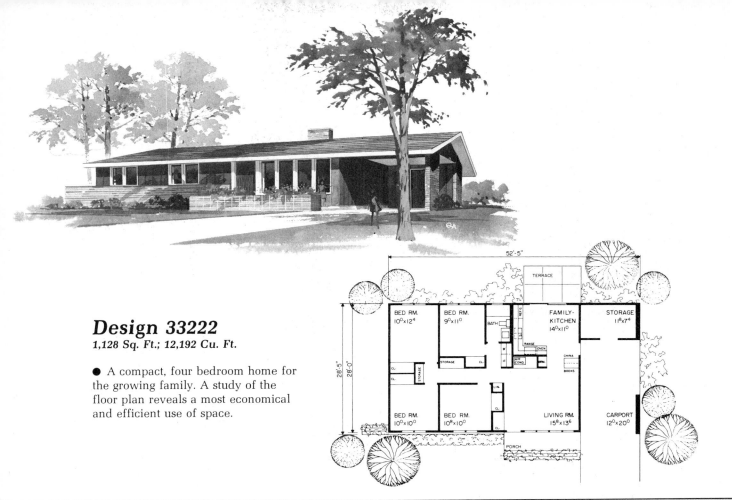

Design 33222
1,128 Sq. Ft.; 12,192 Cu. Ft.

● A compact, four bedroom home for the growing family. A study of the floor plan reveals a most economical and efficient use of space.

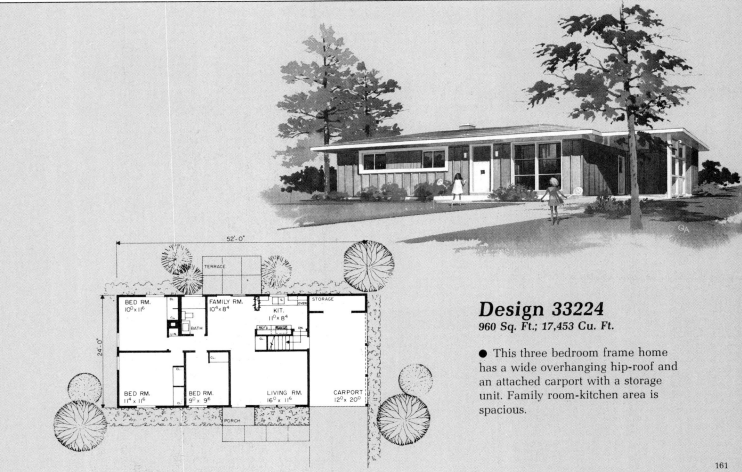

Design 33224
960 Sq. Ft.; 17,453 Cu. Ft.

● This three bedroom frame home has a wide overhanging hip-roof and an attached carport with a storage unit. Family room-kitchen area is spacious.

Design 32153 960 Sq. Ft.; 18,432 Cu. Ft.

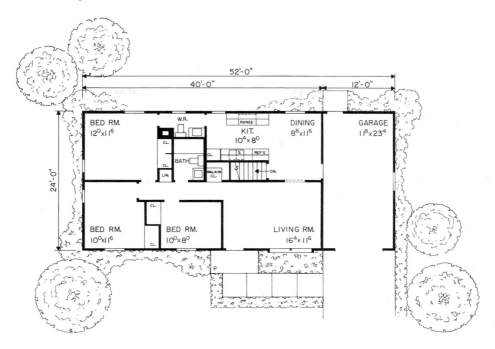

● If you can't make up your mind as to which of the delightful traditional exteriors you like best on the opposing page, you need not decide now. The blueprints you receive show details for the construction of all three front exteriors. However, before you order, decide whether you wish your next home to have a basement or not. If you prefer the basement plan order Design 32153 above. Should your preference be for a non-basement plan you should order blueprints for Design 32154 below. Whatever your choice, you'll forever love the charm of its exterior and the comfort and convenience of the interior. The three bedrooms will serve your family ideally.

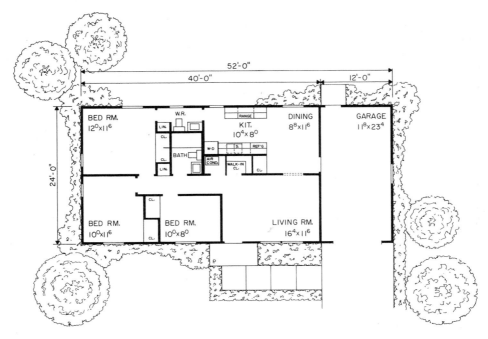

Design 32154 960 Sq. Ft.; 10,675 Cu. Ft.

Design 32194
882 Sq. Ft.; 10,408 Cu. Ft.

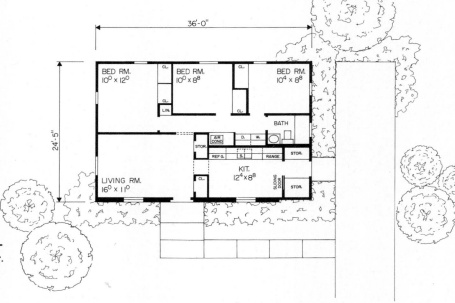

● Here is a basic 36 x 24 foot rectangle that can be built with either the contemporary or traditional exterior. What's your preference? Observe the grouping of plumbing facilities. This is a positive economy factor. The kitchen is sizable and has plenty of space for eating. There will be no lack of space to put things with all the closet and storage areas. The living room has space for effective furniture placement. The washer and dryer are strategically situated at the source of most soiled linen. There is a recessed sliding door between the kitchen and the vestibule.

Design 32195

1,042 Sq. Ft.; 12,840 Cu. Ft. - Non-Basement
19,577 Cu. Ft. - Basement

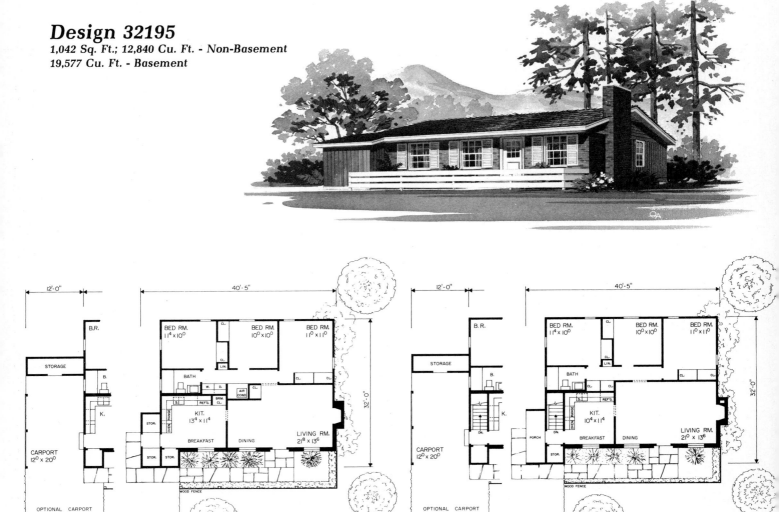

NON BASEMENT PLAN

BASEMENT PLAN

● The construction blueprints you order for this design will feature a number of options. You will enjoy making your decisions on which is your favorite traditional exterior. Then, you and your family will also decide on whether you want to build a basement or non-basement house. Finally, you have the choice of whether or not to have a carport. Three bedrooms, bath, kitchen and living room are standard. Study the plan carefully.

Design 31311 1,050 Sq. Ft.; 11,370 Cu. Ft.

● Delightful design and effective, flexible planning comes in little packages, too. This fine traditional exterior with its covered front entrance features an alternate basement plan. Note how the non-basement layout provides a family room and mud room, while the basement option shows kitchen eating and dining room. Sensible planning.

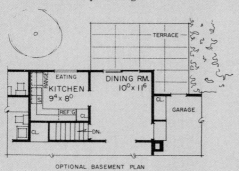

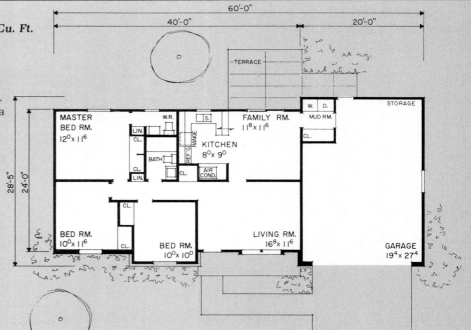

Design 31531
936 Sq. Ft.; 12,252 Cu. Ft.

● This is but another example of delightful custom design applied to a small home. The detailing of the windows and the door, plus the effective use of siding and brick contribute to the charm. The attached garage helps make the house appear even bigger than it really is. The front-to-rear living/dining area opens onto the rear terrace through sliding glass doors. The washer and dryer with wall cabinets above complete the efficient work center area. Observe extra kitchen closet.

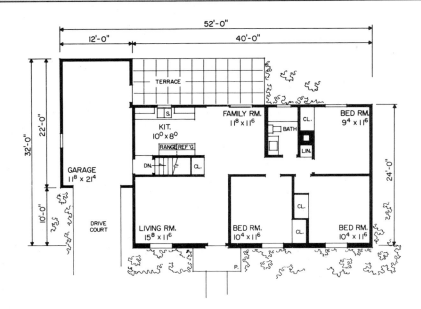

Design 31522
960 Sq. Ft.; 18,077 Cu. Ft.

● Certainly a home to make its occupants proud. The front exterior is all brick veneer, while the remainder of the house and garage is horizontal siding. The slightly overhanging roof, the wood shutters and the carriage lights flanking the front door are among the features that will surely catch the eyes of the passerby. The living room has excellent wall space for furniture placement. The family room, the full basement and the attached garage are other features. Don't miss the sliding glass doors.

167

Design 32198

1,193 Sq. Ft.; 23,263 Cu. Ft. - Basement
14,804 Cu. Ft. - Non-Basement

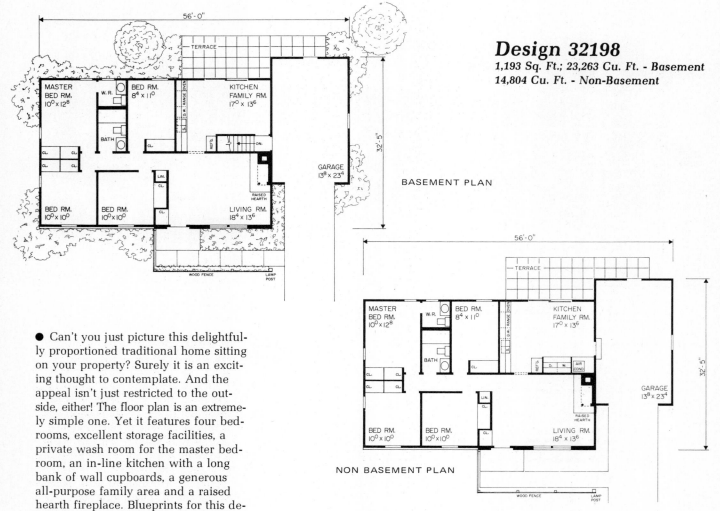

● Can't you just picture this delightfully proportioned traditional home sitting on your property? Surely it is an exciting thought to contemplate. And the appeal isn't just restricted to the outside, either! The floor plan is an extremely simple one. Yet it features four bedrooms, excellent storage facilities, a private wash room for the master bedroom, an in-line kitchen with a long bank of wall cupboards, a generous all-purpose family area and a raised hearth fireplace. Blueprints for this design include both basement and non-basement details for construction.

Design 32199

1,185 Sq. Ft.; 21,721 Cu. Ft. - Basement
13,224 Cu. Ft. - Non-Basement

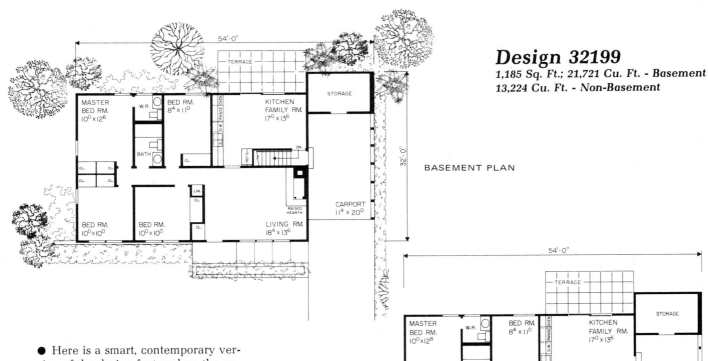

BASEMENT PLAN

NON BASEMENT PLAN

● Here is a smart, contemporary version of the design featured on the opposite page. It has all the amenities for large family livability at modest cost. While the parents have their master bedroom with private wash room, the kids have three bedrooms and are served by the main bath. In addition to the living room, there is extra livability to be enjoyed in the spacious family area which has access to the rear terrace. Don't miss all those closets or the bulk storage room of the carport. Blueprints include optional basement and non-basement details.

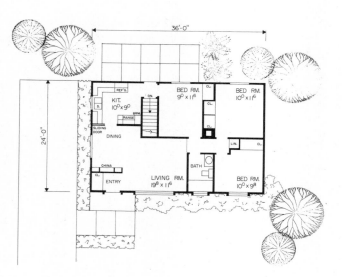

Design 32167
864 Sq. Ft.; 16,554 Cu. Ft.

● This 36' x 24' contemporary rectangle will be economical to build whether you construct the basement design at left, 32167, or the non-basement version below, 32168.

Design 32168
864 Sq. Ft.; 9,244 Cu. Ft.

● This non-basement design features a storage room and a laundry area with cupboards above the washer and dryer. Notice the kitchen eating space.

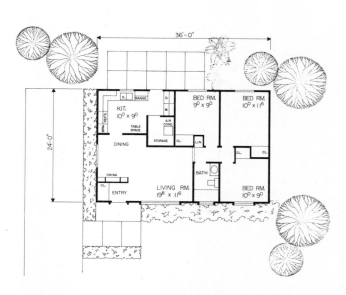

● Here are three optional front exteriors which can be built with either the basement plan, 32160, or the non-basement plan, 32161. Whichever number you order, the blueprints will include details for all three optional front elevations. It is interesting to observe the variations in the two floor plans. Don't miss the extra storage facilities. Notice the location of the stacked washer and dryer in the non-basement plan.

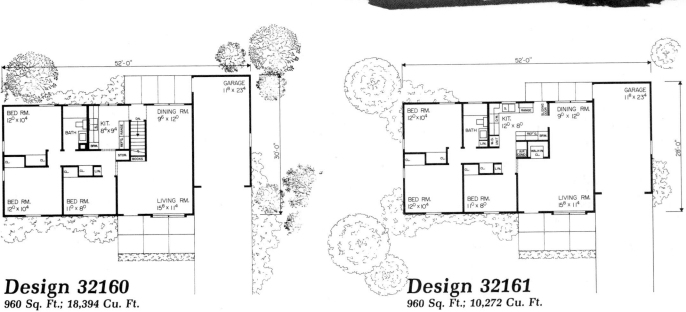

Design 32160
960 Sq. Ft.; 18,394 Cu. Ft.

Design 32161
960 Sq. Ft.; 10,272 Cu. Ft.

Design 31281
1,190 Sq. Ft.; 21,920 Cu. Ft.

● Whatever you call it, contemporary or traditional, or even transitional, the prudent incorporation of newer with old design features results in a pleasant facade. The window and door detailing is obviously from an early era; while the low-pitched roof and the carport are features of relatively more recent vintage. Inside, there is an efficient plan which lends itself to the activities of the small family. As a home for a retired couple, or a young newlywed couple, this will be a fine investment. Certainly it will not require a big, expensive piece of property. The plan offers two full baths, three bedrooms, an excellent L-shaped kitchen, 23-foot living and dining area and a basement. Note side terrace.

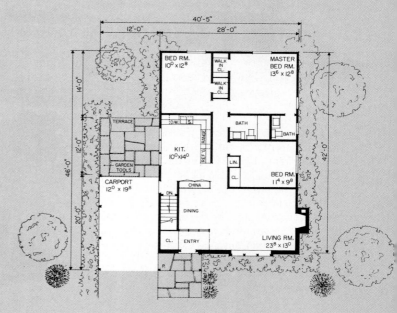

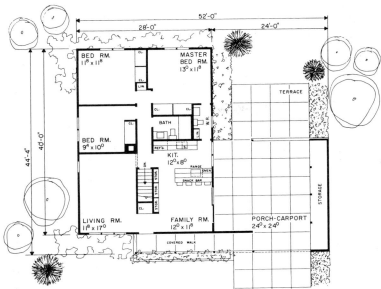

Design 33195
1,120 Sq. Ft.; 20,440 Cu. Ft.

● This 28 x 40 foot rectangle is a fine example of how delightful and refreshing contemporary design can become. Though simple in basic construction and floor planning, this home has an abundance of exterior appeal and interior livability. Observe the low-pitched, wide overhanging roof, the attached 24 foot square covered porch/carport, the effective glass area and vertical siding, plus the raised planter. The center entrance is flanked by the formal living room and the informal, all-purpose family room. A snack bar is accessible from the kitchen. Sliding glass doors lead to the outdoor living area. In addition to the main bath there is an extra wash room handy to both the master bedroom and the kitchen.

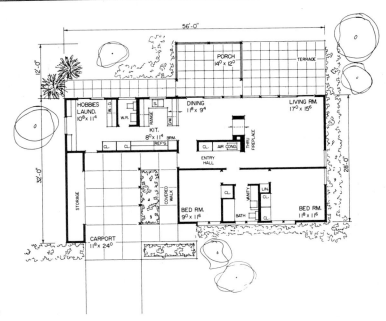

Design 31216
1,184 Sq. Ft.; 10,656 Cu. Ft.

● Whether called upon to function as a home for a retired couple or for a small family, this frame, non-basement design will provide wonderful livability. The covered walk with its adjacent planting areas makes for a pleasing approach to the front door. The thru-fireplace may be enjoyed from both formal areas—the living room and the separate dining room. Glass sliding doors lead to the cheerful enclosed porch. A pass-thru from the kitchen facilitates the serving of meals in the dining area. A wash room is conveniently located between the kitchen and hobbies/laundry room and is handy from the outdoors. The general area has an abundance of counter, cupboard and closet space.

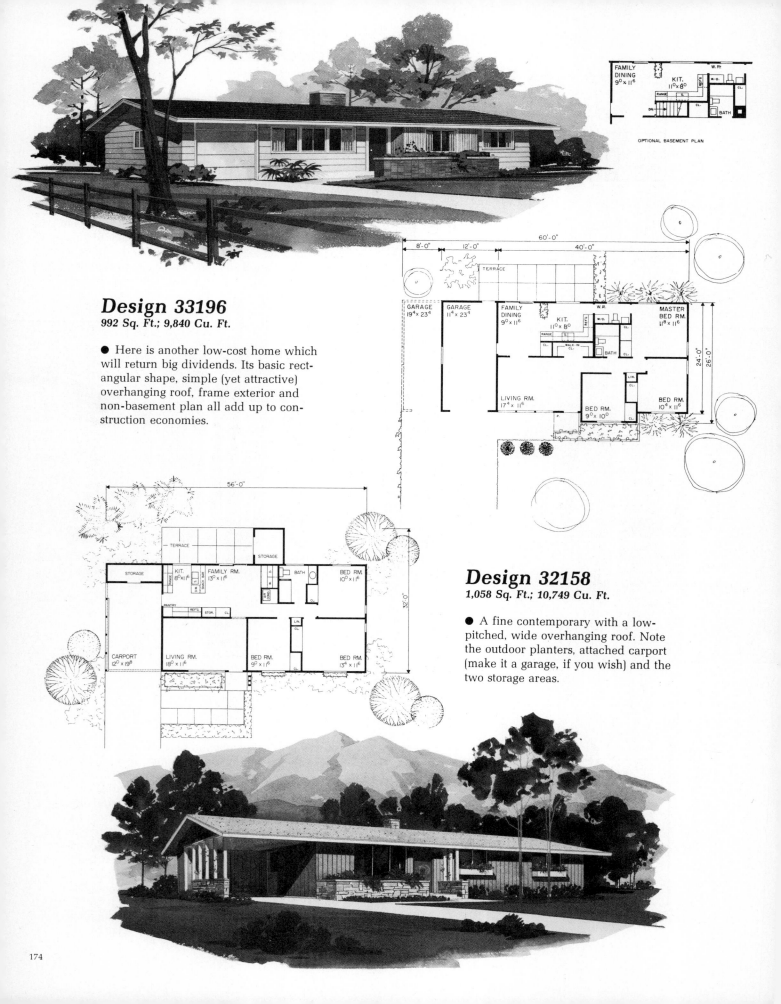

Design 33196
992 Sq. Ft.; 9,840 Cu. Ft.

● Here is another low-cost home which will return big dividends. Its basic rectangular shape, simple (yet attractive) overhanging roof, frame exterior and non-basement plan all add up to construction economies.

Design 32158
1,058 Sq. Ft.; 10,749 Cu. Ft.

● A fine contemporary with a low-pitched, wide overhanging roof. Note the outdoor planters, attached carport (make it a garage, if you wish) and the two storage areas.

Design 33226
1,113 Sq. Ft.; 11,760 Cu. Ft.

● Three bedrooms, 1½ baths, a laundry area, an all-purpose family room and an L-shaped kitchen with eating space are among the features of this traditional. Back-to-back plumbing will be very economical.

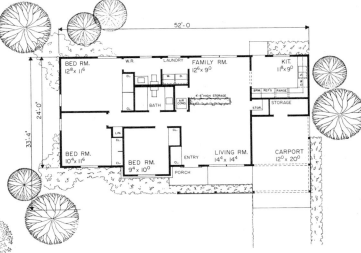

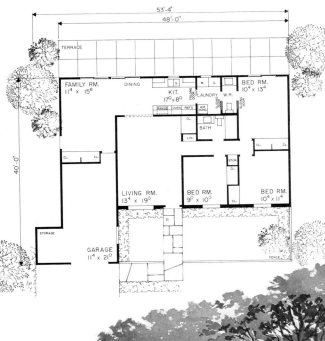

Design 33227
1,164 Sq. Ft.; 13,479 Cu. Ft.

● This efficient plan contains all the elements for a lifetime of living convenience. Of particular interest is the economical grouping of all plumbing facilities at the rear of the plan.

175

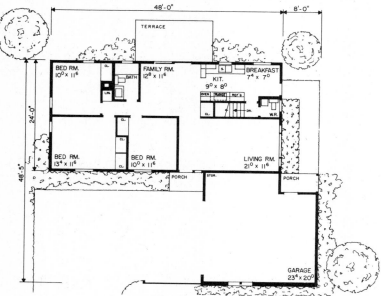

Design 33208
1,152 Sq. Ft.; 21,888 Cu. Ft.

● This appealing traditional, L-shaped home has much to offer those in search of a moderately sized home which can be built within the confines of a relatively small budget. First of all, consider the charm of the exterior. Surely this will be one of the most appealing houses on the street. And little wonder. Its proportion and architectural detailing are excellent. The shutters, the window treatment, the roof lines, the planter and the fence and lamp post are all captivating features.

Design 31373
1,200 Sq. Ft.; 15,890 Cu. Ft.

● A traditional L-shaped home with an attractive recessed front entrance which leads into a floor plan where traffic patterns are most efficient. It is possible to go from one room to any other without needlessly walking through a third room. The daily household chores will be easily dispatched. The U-shaped kitchen has an abundance of cupboard and counter space, plus a pass-thru to the snack bar in the family room. The washer and dryer location is handy.

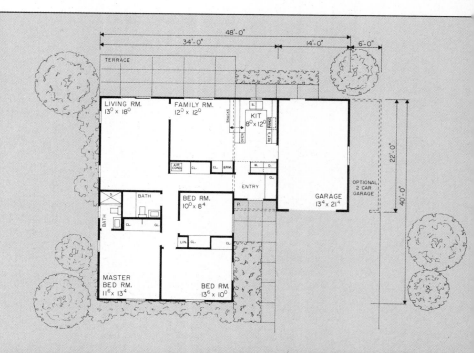

Design 31058
1,200 Sq. Ft.; 13,392 Cu. Ft.

● When you build this charming L-shaped traditionally designed home don't leave out the wood fence and lamp post. These are just the features needed to complete the picture. A porch shelters the front door which leads to the centered entry flanked by the formal living room and the informal family room. The kitchen is but a step from the separate dining room. The laundry equipment, the extra wash room and heater location are all grouped together.

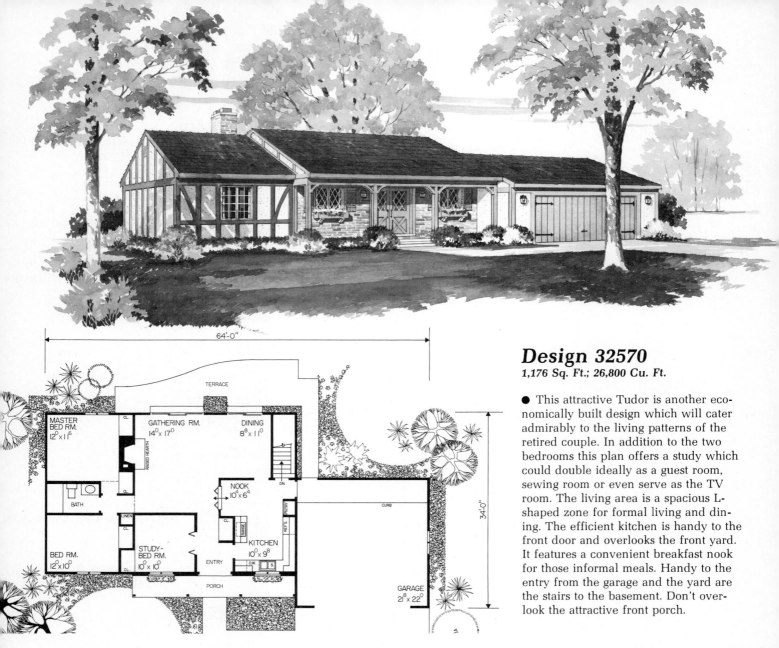

Design 32570
1,176 Sq. Ft.; 26,800 Cu. Ft.

● This attractive Tudor is another economically built design which will cater admirably to the living patterns of the retired couple. In addition to the two bedrooms this plan offers a study which could double ideally as a guest room, sewing room or even serve as the TV room. The living area is a spacious L-shaped zone for formal living and dining. The efficient kitchen is handy to the front door and overlooks the front yard. It features a convenient breakfast nook for those informal meals. Handy to the entry from the garage and the yard are the stairs to the basement. Don't overlook the attractive front porch.

Design 32607
1,208 Sq. Ft.; 15,183 Cu. Ft.

● Here is an English Tudor retirement cottage. Its byword is "convenience". There are two sizable bedrooms, a full bath, plus an extra wash room. The living and dining areas are spacious and overlook both front and rear yards. Sliding glass doors in both these areas lead to the outdoor terrace. Note the fireplace in the living room. In addition to the formal dining area with its built-in china cabinet, there is a delightful breakfast eating alcove in the kitchen. The U-shaped work area is wonderfully efficient. The laundry is around the corner. Blueprints include optional basement details.

OPTIONAL BASEMENT

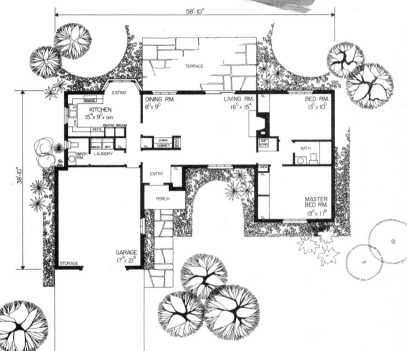

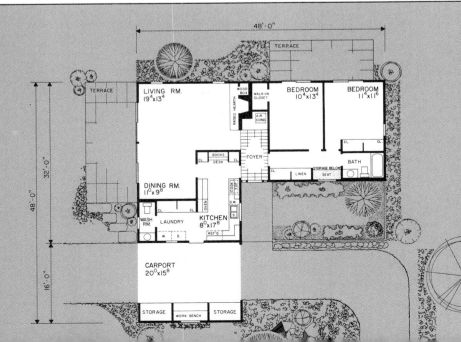

Design 31279
1,200 Sq. Ft.; 12,300 Cu. Ft.

● This cozy traditional ranch house is ideal for a retired couple. Its small size and one-story plan make it easy for them to enjoy and maintain. Its low cost also makes it feasible for young married couples, who, as their families and income expands, can add bedrooms to the front or side of the bedroom wing. The house provides good circulation to living, sleeping and utility zones so that no room has the interruption of cross traffic.

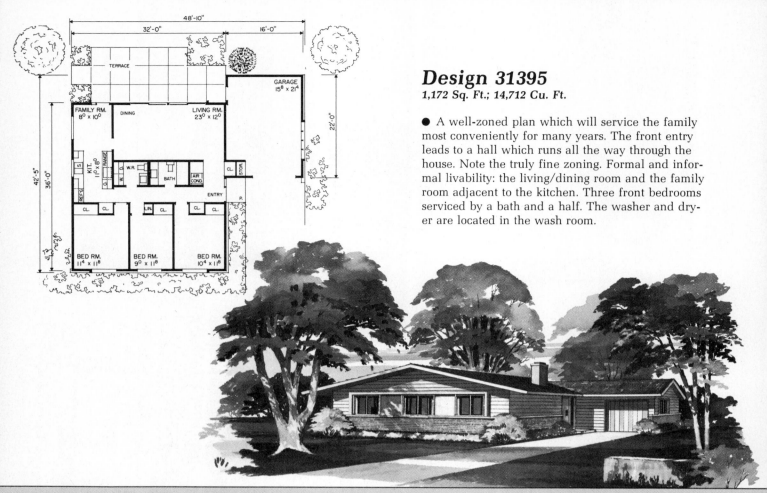

Design 31395
1,172 Sq. Ft.; 14,712 Cu. Ft.

● A well-zoned plan which will service the family most conveniently for many years. The front entry leads to a hall which runs all the way through the house. Note the truly fine zoning. Formal and informal livability: the living/dining room and the family room adjacent to the kitchen. Three front bedrooms serviced by a bath and a half. The washer and dryer are located in the wash room.

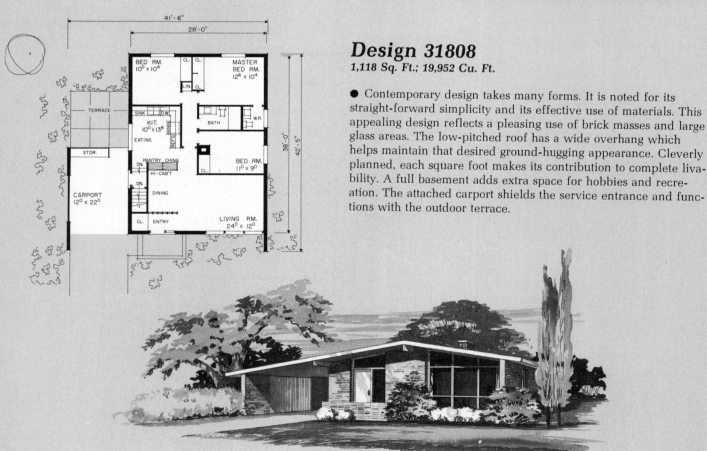

Design 31808
1,118 Sq. Ft.; 19,952 Cu. Ft.

● Contemporary design takes many forms. It is noted for its straight-forward simplicity and its effective use of materials. This appealing design reflects a pleasing use of brick masses and large glass areas. The low-pitched roof has a wide overhang which helps maintain that desired ground-hugging appearance. Cleverly planned, each square foot makes its contribution to complete livability. A full basement adds extra space for hobbies and recreation. The attached carport shields the service entrance and functions with the outdoor terrace.

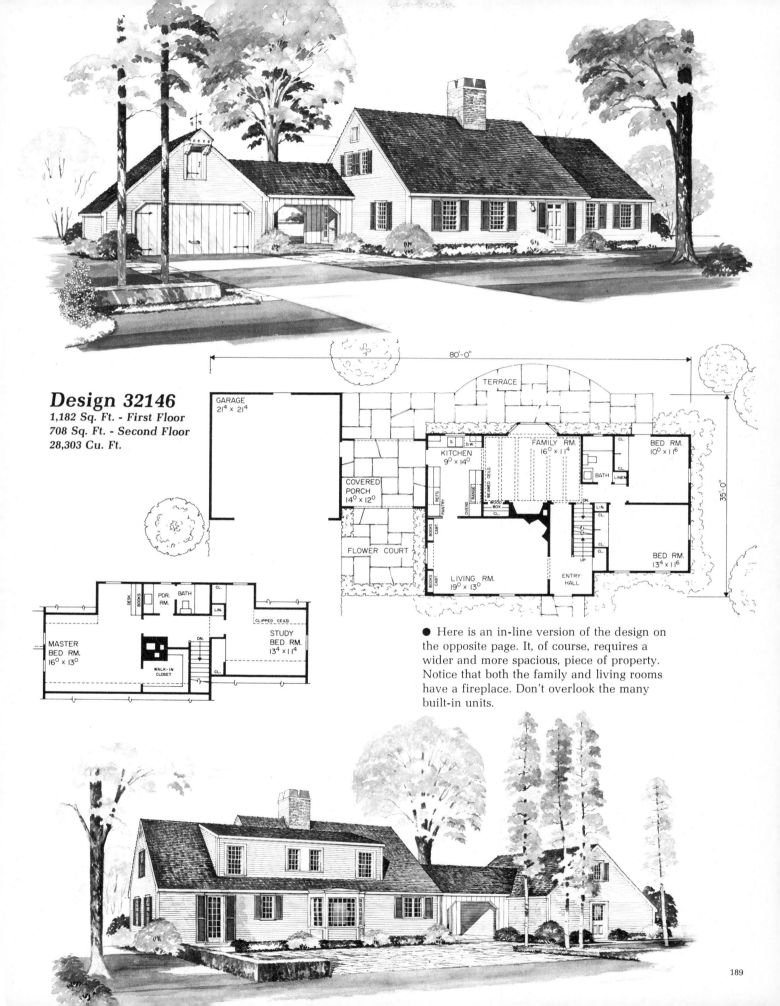

Design 32146

1,182 Sq. Ft. - First Floor
708 Sq. Ft. - Second Floor
28,303 Cu. Ft.

● Here is an in-line version of the design on the opposite page. It, of course, requires a wider and more spacious, piece of property. Notice that both the family and living rooms have a fireplace. Don't overlook the many built-in units.

Design 31084

1,804 Sq. Ft. - First Floor
732 Sq. Ft. - Second Floor
33,842 Cu. Ft.

● A terrific contemporary design. Two bedrooms, 1½ baths, formal and informal living available in the main living unit of the first floor. Two more bedrooms plus two full baths on the second floor if developed. The family room's features include a wall of built-ins, fireplace and sliding glass doors to the terrace.

Design 33197
1,434 Sq. Ft. - First Floor
750 Sq. Ft. - Second Floor; 31,286 Cu. Ft.

● This contemporary one-and-a-half story will offer a refreshing change of pace to any neighborhood. Its appealing character will elicit a pleasing comment from friends and strangers alike. Like so many houses of this type, this design may be finished off in stages. If your family is small now, you may wish to postpone the completion of the second floor until a later date. There are still two bedrooms and a full bath downstairs.

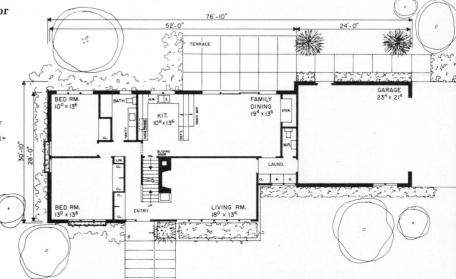

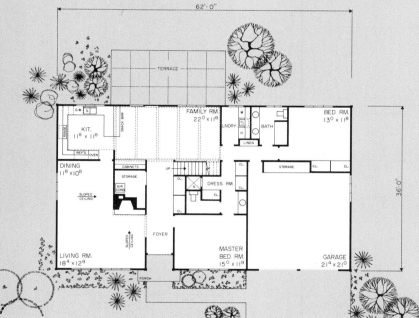

Design 32309
1,719 Sq. Ft. - First Floor
456 Sq. Ft. - Second Floor; 22,200 Cu. Ft.

● Here's proof that the simple rectangle (which is relatively economical to build, naturally) can, when properly planned, result in unique living patterns. The exterior can be exceedingly appealing, too. Study the floor plan carefully. The efficiency of the kitchen could hardly be improved upon. It is strategically located to serve the formal dining room, the family room and even the rear terrace. The sleeping facilities are arranged so the second floor can be developed at a later date.

Design 32278 1,804 Sq. Ft. - First Floor; 939 Sq. Ft. - First Floor; 44,274 Cu. Ft.

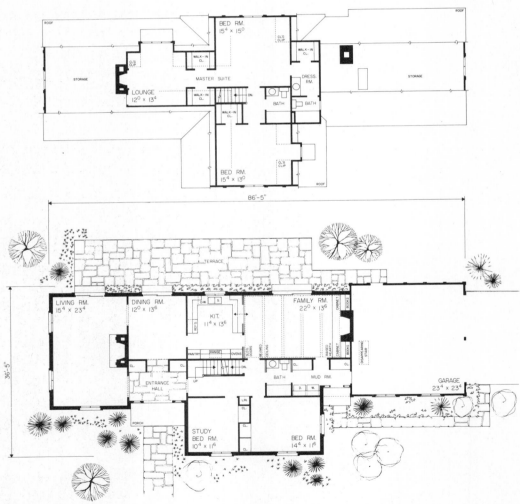

● This cozy Tudor adaptation is surely inviting. Its friendly demeanor seems to say, "welcome". Upon admittance to the formal front entrance hall, even the most casual of visitors will be filled with anticipation at the prospect of touring the house. And little wonder, too. Traffic patterns are efficient. Room relationships are excellent. A great feature is the location of the living, dining, kitchen and family rooms across the back of the house. Each enjoys a view of the rear yard and sliding glass doors provide direct access to the terrace. Another outstanding feature is the flexibility of the sleeping patterns. This may be a five bedroom house, or one with three bedrooms with study and lounge. Don't miss the three fireplaces and three baths.

● Features are plenty - both inside and out. A list of the exterior design highlights is most interesting. It begins with the character created by the impressive roof surfaces. The U-shape creates a unique appeal and results in the formation of a garden court. The covered passage provides the court with its complete privacy from the street. The detailing of the facades of the projecting wings are delightful, indeed. The massive chimney is worthy of particular note. The living room features a sloping beamed ceiling. It also has space left over for the formal dining facilities. The family room has a flat beamed ceiling. The sleeping facilities are exceptional.

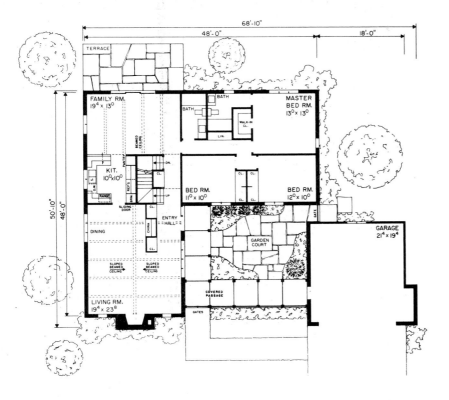

Design 32127
1,712 Sq. Ft. - First Floor
450 Sq. Ft. - Second Floor
39,435 Cu. Ft.

Design 31904

1,760 Sq. Ft. - First Floor
900 Sq. Ft. - Second Floor; 42,615 Cu. Ft.

● You'll certainly have fun living in this traditional one-and-a-half story design. If your family is of modest size you'll have many options as to how you may wish to use some of the rooms. The first floor bedrooms can house two or three children, leaving the whole upstairs for the parents. Or, one or two children might like to locate on the first floor with a hobby room next door. Whatever the arrangement the parents will have their lounge with fireplace. Study the remainder of the plan for your favorite features.

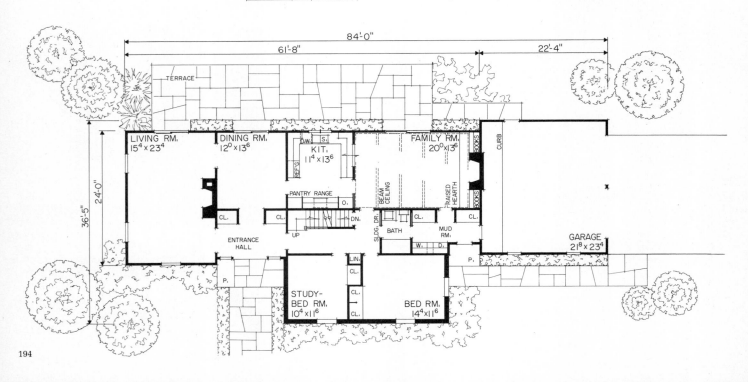

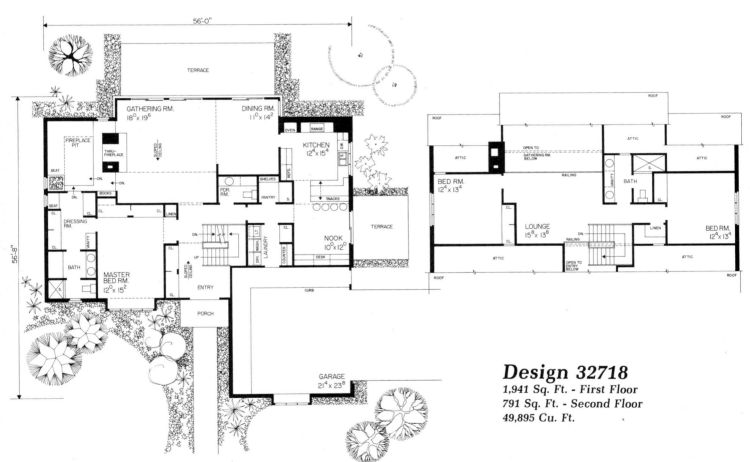

Design 32718

1,941 Sq. Ft. - First Floor
791 Sq. Ft. - Second Floor
49,895 Cu. Ft.

● You and your family will just love the new living patterns you'll experience in this story-and-a-half. The front entry hall features twin coat closets and an impressive open staircase to the upstairs and basement. The master bedroom has a compartmented bath with both tub and stall shower. There is the dressing room with plenty of storage which steps down into a unique, sunken conversation pit. This cozy area has a planter, built-in seat and a view of the thru-fireplace, opening to the gathering room as well. Here, the ceiling slopes to the top of the second floor lounge which looks down into the gathering room. The efficient kitchen effectively serves the formal dining room and informal nook. Both the younger and older generations will love the privacy this design offers.

Design 31115

1,440 Sq. Ft. - First Floor
740 Sq. Ft. - Second Floor
33,516 Cu. Ft.

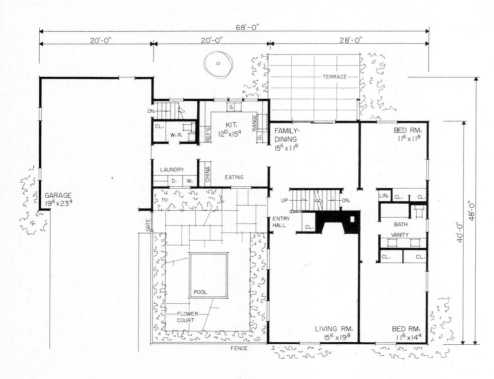

● A most distinctive exterior with an equally distinctive interior. A host of architectural features highlight the eye-catching facade. Note the flower court with small pool. Surely an impressive welcome to this country style home. A study of the plan reveals all the elements to assure convenient living. The main living unit, the first floor, functions very efficiently. Two bedrooms and a full bath comprise the sleeping zone. Formal, front living room with fireplace. The U-shaped kitchen is very efficient. A built-in china cabinet is in the eating area of the kitchen. The family/dining room will serve the family admirably. It has sliding glass doors to the rear terrace. Adjacent to the kitchen is the laundry area, wash room and entrance from the garage. This living unit is definitely complete. Now add the second floor. Absolutely fantastic! The whole second floor is a master bedroom suite consisting of a bedroom, full bath, dressing room and lounge with alcove. If you so desire, the lounge could instead be the fourth bedroom if it is needed.

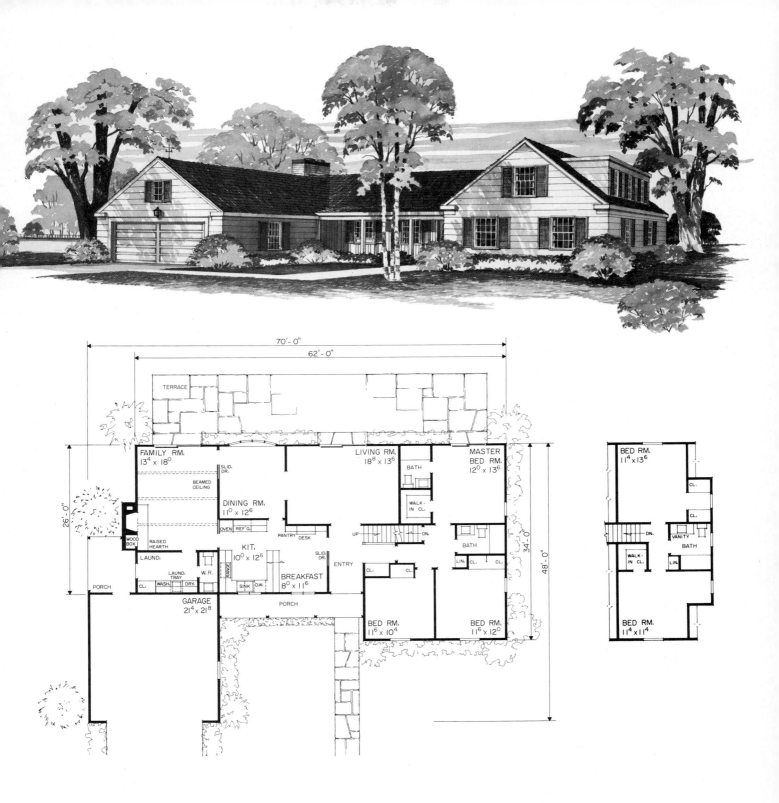

Design 31967 1,804 Sq. Ft. - First Floor; 496 Sq. Ft. - Second Floor; 40,173 Cu. Ft.

● You'll always want that first impression your guests get of your new home to be a lasting one. There will be much that will linger in the memories of most of your visitors after their visit to this home. Of course, the impressive exterior will long be remembered. And little wonder with its distinctive projecting garage and bedroom wing, its recessed front porch, its cleanly appearing horizontal siding and its interesting roof lines. Inside, there is much to behold. The presence of five bedrooms and three full baths will not be forgotten soon. Nor will the arrangement of the living zone - the family, dining and living room - which overlooks and functions with the rear yard.

Design 31790
1,782 Sq. Ft. - First Floor
920 Sq. Ft. - Second Floor
37,359 Cu. Ft.

● A versatile plan wrapped in a pleasing traditional facade to cater to the demands of even the most active of families. There is plenty of living space for both formal and informal activities. Locating two bedrooms upstairs and two down, sleeping accommodations are excellent. In addition to the two full baths of the second floor, there is a full bath plus an extra wash room on the first. As for closets, cabinets, cuboards and shelves, they can be found "all over the place". Also worthy of note is the raised hearth fireplace in the family room, the snack bar and pass-thru to kitchen. The first floor laundry is standard even though there is a basement included. Surely an impressive home which will please the family for many years to come.

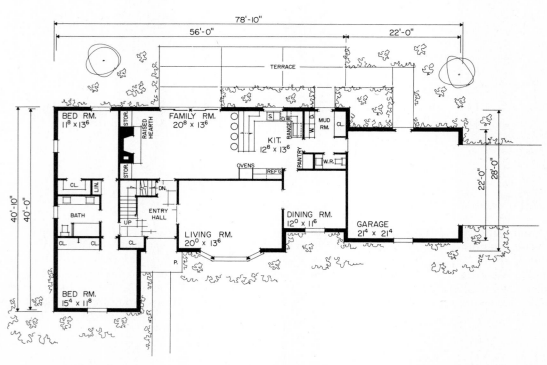

Design 31793
1,986 Sq. Ft. - First Floor
944 Sq. Ft. - Second Floor
35,800 Cu. Ft.

● How's this for traditional charm? The large bay windows flanking the recessed front entrance with its double doors are delightful features. The twin dormers above are attractive, too. The orientation of the work center is interesting. The U-shape kitchen overlooks the front yard and, along with mud room and adjacent wash room, is only a few steps from the front service entrance, the door to garage and access to the rear terrace. The kitchen features many built-ins, a separate dining area, pantry and snack bar. The family room has a raised hearth fireplace, wood box and sliding glass doors to the terrace. Of particular interest are the sleeping facilities. Upstairs there are three nice sized bedrooms, walk-in closets, a dressing room and two baths. Downstairs there are two bedrooms and a full bath which function well together and could comprise a separate suite.

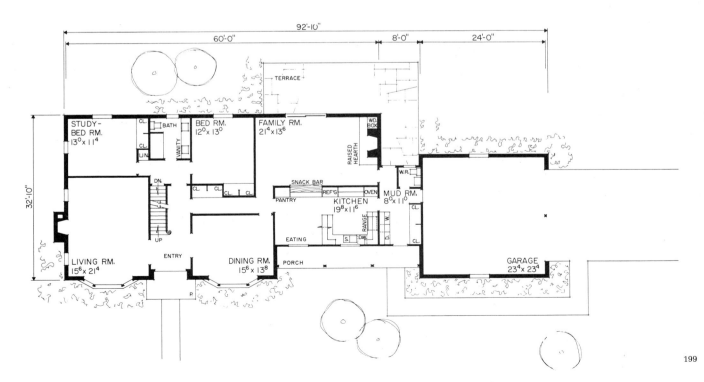

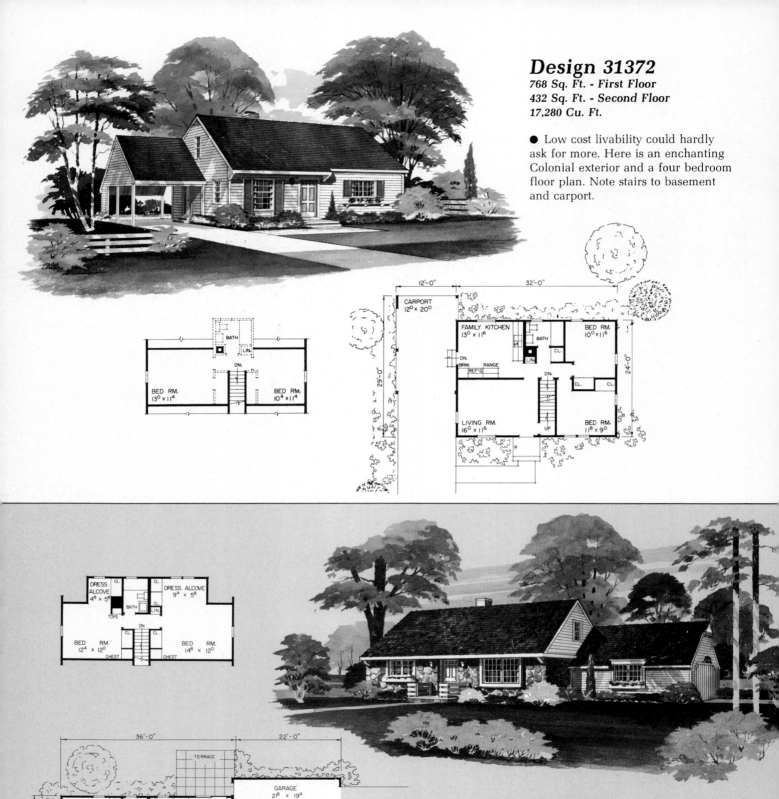

Design 31372
768 Sq. Ft. - First Floor
432 Sq. Ft. - Second Floor
17,280 Cu. Ft.

● Low cost livability could hardly ask for more. Here is an enchanting Colonial exterior and a four bedroom floor plan. Note stairs to basement and carport.

Design 33189
884 Sq. Ft. - First Floor
598 Sq. Ft. - Second Floor; 18,746 Cu. Ft.

● Four bedrooms, two baths, a large kitchen/dining area, plenty of closets, a full basement and an attached two-car garage are among the highlights of this design. Note the uniqueness of the second floor.

Design 31394

832 Sq. Ft. - First Floor
512 Sq. Ft. - Second Floor
19,385 Cu. Ft.

● The growing family with a restricted building budget will find this a great investment - a convenient living floor plan inside an attractively designed facade.

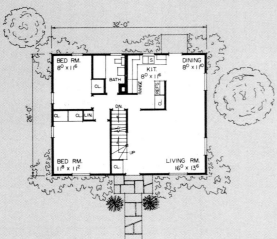

Design 32510

1,191 Sq. Ft. - First Floor
533 Sq. Ft. - Second Floor
27,500 Cu. Ft.

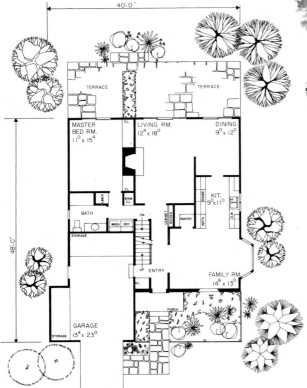

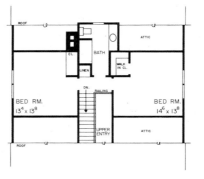

● The pleasant in-line kitchen is flanked by a separate dining room and a family room. The master bedroom is on the first floor with two more bedrooms upstairs.

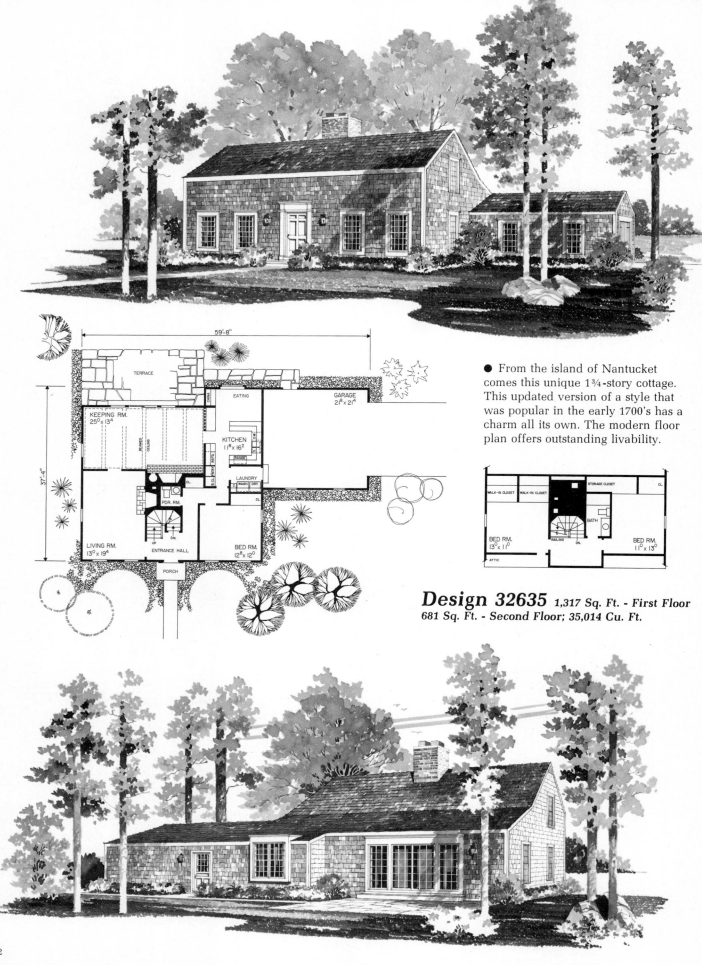

● From the island of Nantucket comes this unique 1¾-story cottage. This updated version of a style that was popular in the early 1700's has a charm all its own. The modern floor plan offers outstanding livability.

Design 32635 1,317 Sq. Ft. - First Floor
681 Sq. Ft. - Second Floor; 35,014 Cu. Ft.

● Another 1¾-story home - a type of house favored by many of Cape Cod's early whalers. The compact floor plan will be economical to build and surely an energy saver. An excellent house to finish-off in stages.

Design 32636 1,211 Sq. Ft. - First Floor
747 Sq. Ft. - Second Floor; 28,681 Cu. Ft.

Design 33131
1,166 Sq. Ft. - First Floor
709 Sq. Ft. - Second Floor; 29,328 Cu. Ft.

● A charming traditional adaptation with an extra measure of exterior appeal and an outstanding array of interior features. Contributing to the beauty of the exterior is the prudent use of varying materials. There are cedar shakes, quarried stone and both horizontal and vertical siding. The covered porch sheltering the front door and the large living room window is a highlight. Inside, there is a fine functioning floor plan. There are four bedrooms, two full baths, both a formal and an informal dining area, an excellent kitchen, a sizable living room with fireplace, plenty of closets and a basement. This design would function fine as a three bedroom home. The breakfast room and rear bedroom could combine to make a family room.

NOTE: For those interested in reviewing a larger selection of 1½ and two-story homes, a 320 page plan book is available entitled *400 1½ and Two-Story Designs.* Your remittance in the amount of $5.95 per copy can be sent to Home Planners, Inc., Dept. CV, 23761 Research Drive, Farmington Hills, Michigan 48024.

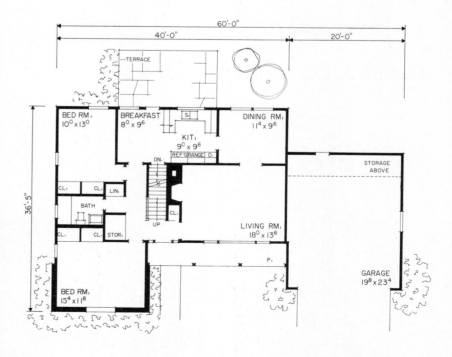

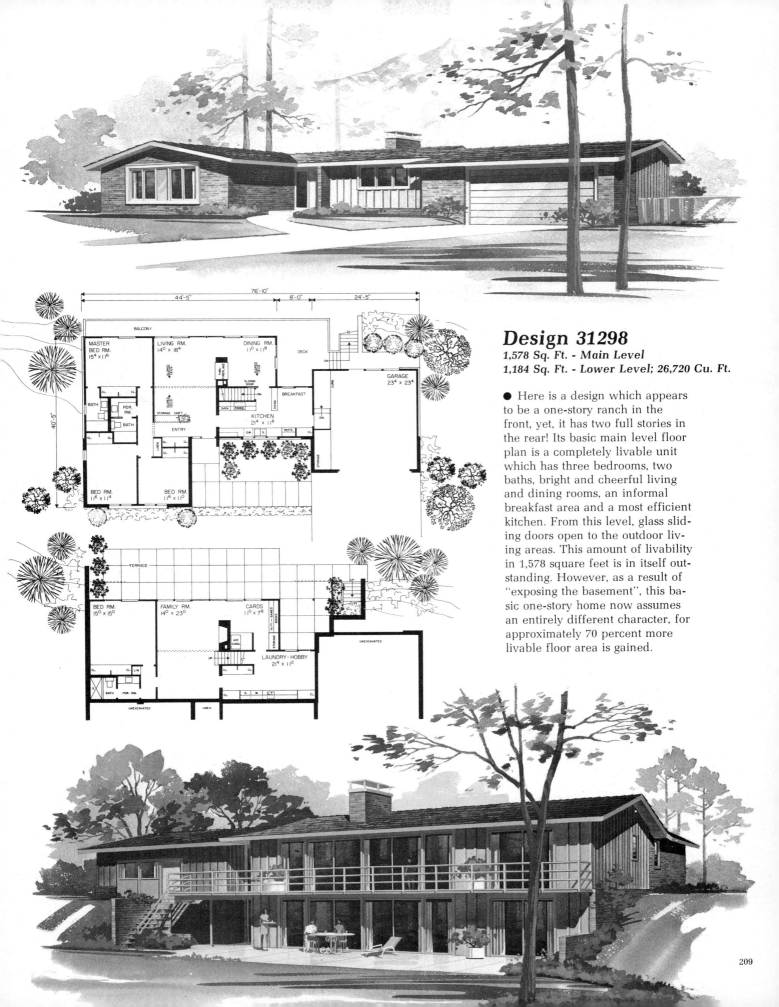

Design 31298
1,578 Sq. Ft. - Main Level
1,184 Sq. Ft. - Lower Level; 26,720 Cu. Ft.

● Here is a design which appears to be a one-story ranch in the front, yet, it has two full stories in the rear! Its basic main level floor plan is a completely livable unit which has three bedrooms, two baths, bright and cheerful living and dining rooms, an informal breakfast area and a most efficient kitchen. From this level, glass sliding doors open to the outdoor living areas. This amount of livability in 1,578 square feet is in itself outstanding. However, as a result of "exposing the basement", this basic one-story home now assumes an entirely different character, for approximately 70 percent more livable floor area is gained.

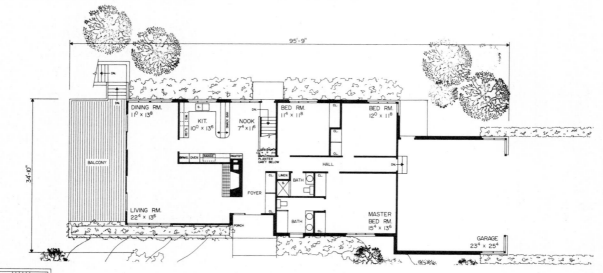

Design 32272

1,731 Sq. Ft. - Main Level
672 Sq. Ft. - Lower Level; 27,802 Cu. Ft.

● Certainly not a huge house. But one, nevertheless, that is long on livability and one that will surely be fun to live in. With its wide-overhanging hip roof, this unadorned facade is the picture of simplicity. As such, it has a quiet appeal all its own. The living-dining area is one of the focal points of the plan. It is wonderfully spacious. The large glass areas and the accessibility through sliding glass doors of the outdoor balcony are fine features. For recreation, there is the lower level which opens onto the covered terrace. The laundry area is also located here.

Design 32504   1,918 Sq. Ft. - Main Level; 1,910 Sq. Ft. - Lower Level; 39,800 Cu. Ft.

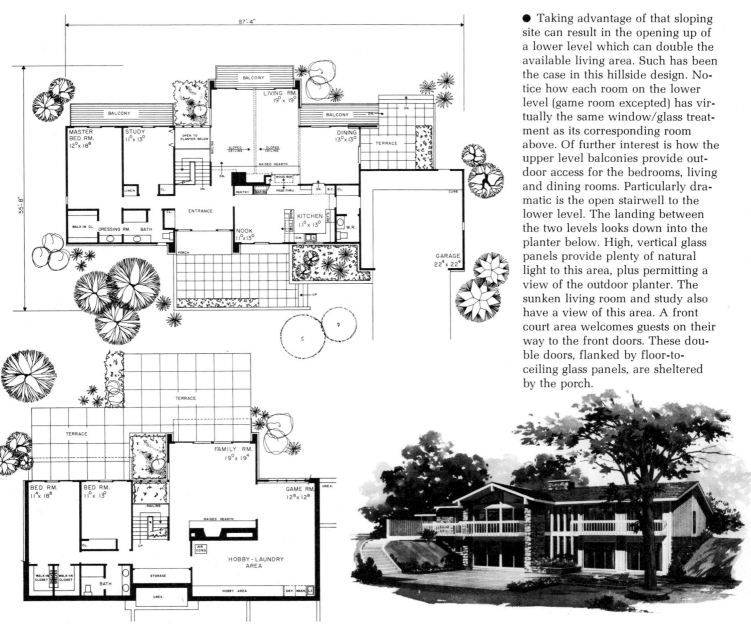

● Taking advantage of that sloping site can result in the opening up of a lower level which can double the available living area. Such has been the case in this hillside design. Notice how each room on the lower level (game room excepted) has virtually the same window/glass treatment as its corresponding room above. Of further interest is how the upper level balconies provide outdoor access for the bedrooms, living and dining rooms. Particularly dramatic is the open stairwell to the lower level. The landing between the two levels looks down into the planter below. High, vertical glass panels provide plenty of natural light to this area, plus permitting a view of the outdoor planter. The sunken living room and study also have a view of this area. A front court area welcomes guests on their way to the front doors. These double doors, flanked by floor-to-ceiling glass panels, are sheltered by the porch.

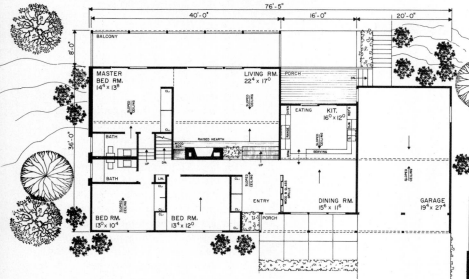

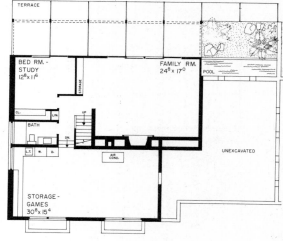

Design 31714
1,808 Sq. Ft. - Main Level
800 Sq. Ft. - Lower Level
33,160 Cu. Ft.

● This hillside design has fine zoning to help assure outstanding living patterns for the active family with a wide variety of interests. In addition to the main living level, there is the bonus of the lower level - a big, multi-purpose area. The unique plan locates the quiet living room and the master bedroom together. These two rooms have their own covered porch, or balcony, which looks down into the rear yard. The children's bedrooms, along with the dining room and kitchen, are also situated together. Just outside the children's bedrooms are the stairs to the lower level. Here is found a fourth bedroom (or study) and the informal family room. This level functions ideally with its terrace.

Design 31976 1,616 Sq. Ft. - Main Level; 1,472 Sq. Ft. - Lower Level; 29,909 Cu. Ft.

● Here's a hillside design just patterned for the large, active family. Whatever the pursuits and interests of the various members, you'd have to guess there would be more than enough space to service one and all with plenty of room to spare. If the children were teenagers, just imagine the fun they would have with their bedrooms, their family room and their hobby room on the lower level. The parents would be equally thrilled with their adult facilities. A home to be enjoyed by all.

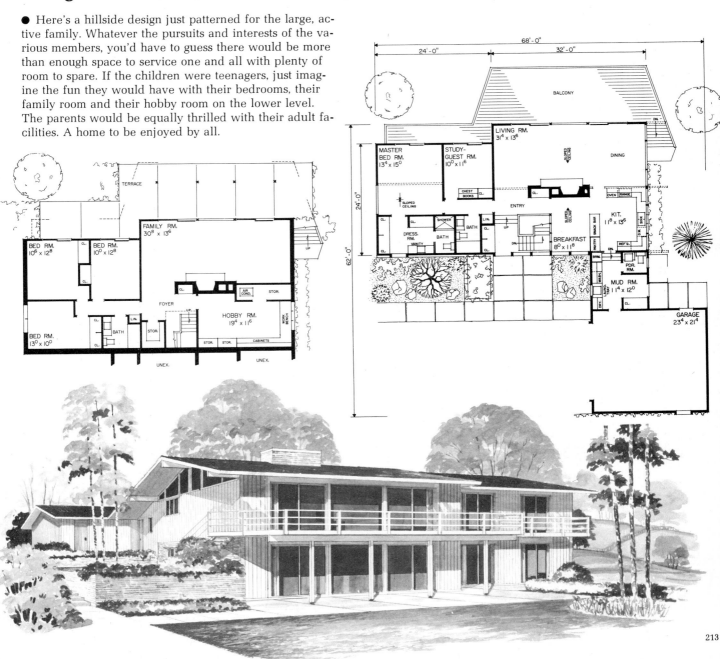

213

Design 31812

1,726 Sq. Ft. - Main Level
1,320 Sq. Ft. - Lower Level
30,142 Cu. Ft.

● A home with two faces. The street view of this contemporary design presents a most pleasingly formal facade. The wide overhanging roof, the projecting masonry piers, the attractive glass treatment and the recessed front entrance all go together to form a perfectly delightful image. The rear terrace view is a picture of informality.

The upper level outdoor balcony overhanging the sweeping lower level terrace will make summer entertaining an occasion to look forward to. The balcony gives way to the spacious deck which will be an ideal spot on which to sun-bathe or dine. The plan is interesting, indeed. A two-way fireplace separates the big living/dining area. Observe the work center; study the lower level layout. Be sure to note the extra bedroom.

NOTE: Should you wish to review an additional selection of split-levels, bi-levels and hillside houses and plans, you may order *205 Multi-Level Designs* at $3.95 per copy from Home Planners, Inc., Dept. CV, 23761 Research Drive, Farmington Hills, Michigan 48024.

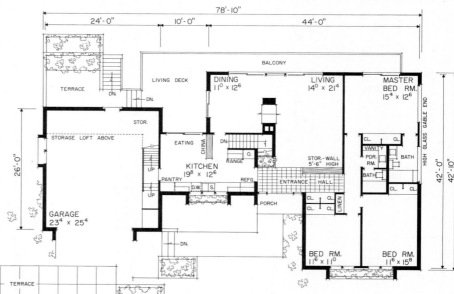

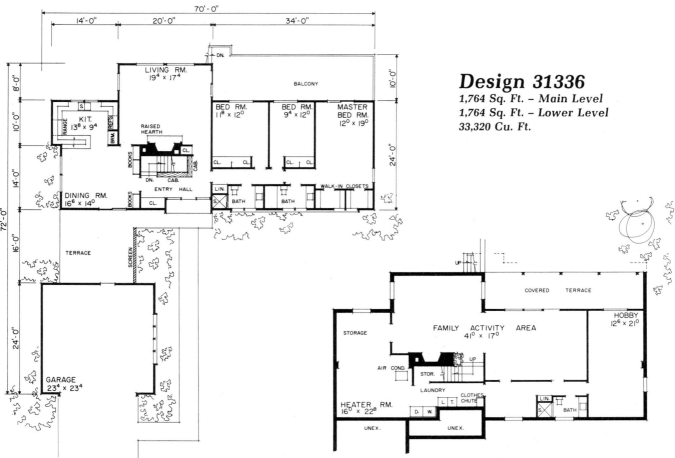

Design 31336

1,764 Sq. Ft. – Main Level
1,764 Sq. Ft. – Lower Level
33,320 Cu. Ft.

● This hillside home is a fine example of what can be done to intergrate both house and site so that they function together effectively. In this case the lot drops off sharply at the rear giving the front the appearance of a one-story ranch. From the rear a partial two-story appearance results. Note how living areas function toward the rear and enjoy to the fullest a commanding view. The sleeping wing features two full baths, wonderful wardrobe storage space and sliding glass doors from the master bedroom to the large outdoor balcony. This balcony results in providing the terrace of the lower level with cover - a welcome feature when those sudden summer showers hit during a family barbeque. The family activity level offers great potential for future development and its spaciousness is just right for those relaxed evenings by the fireplace. A contemporary home to be appreciated by all.

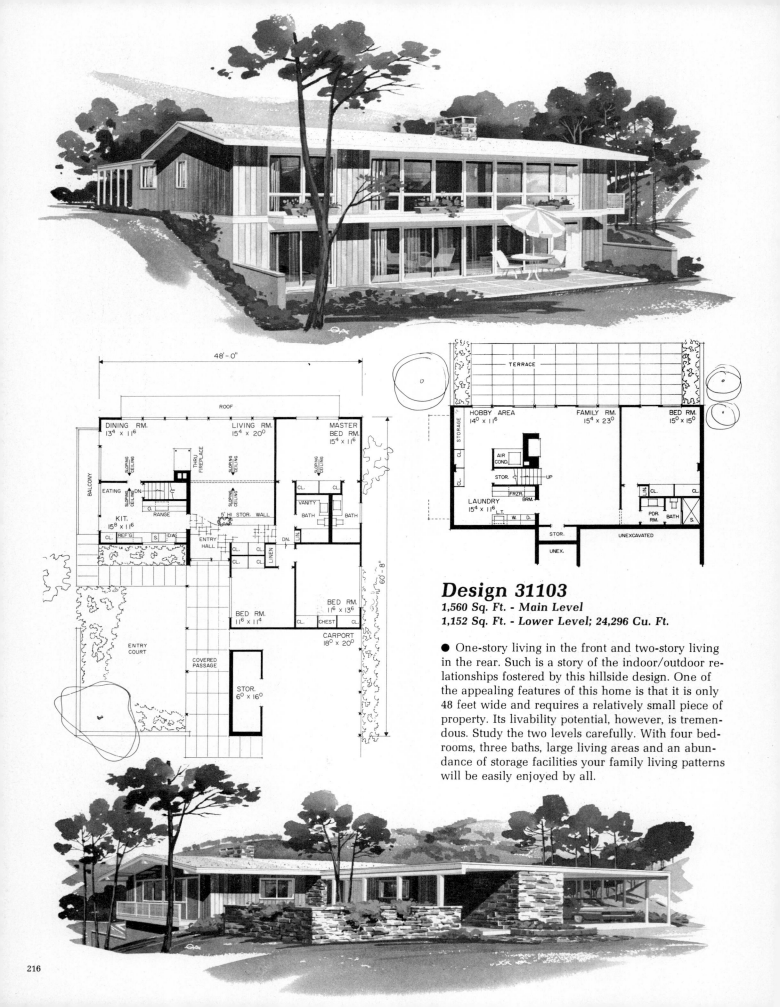

Vacation Homes
For Leisure Living Lifestyles

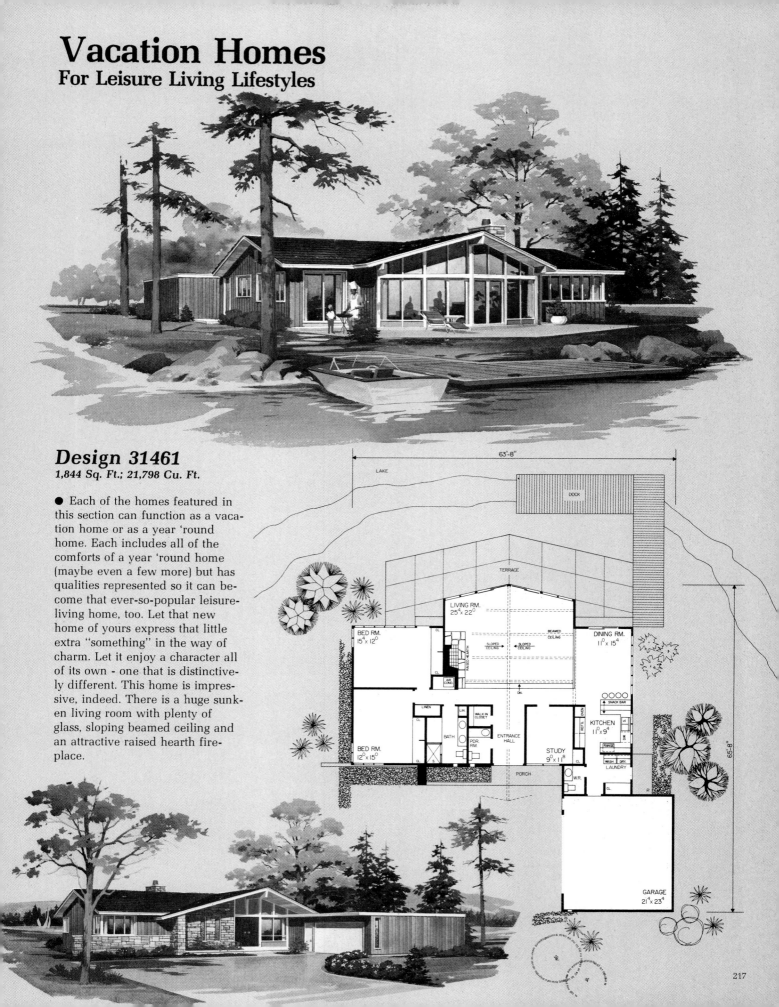

Design 31461
1,844 Sq. Ft.; 21,798 Cu. Ft.

● Each of the homes featured in this section can function as a vacation home or as a year 'round home. Each includes all of the comforts of a year 'round home (maybe even a few more) but has qualities represented so it can become that ever-so-popular leisure-living home, too. Let that new home of yours express that little extra "something" in the way of charm. Let it enjoy a character all of its own - one that is distinctively different. This home is impressive, indeed. There is a huge sunken living room with plenty of glass, sloping beamed ceiling and an attractive raised hearth fireplace.

Design 31497
1,292 Sq. Ft.; 14,274 Cu. Ft.

● Another design whose general shape is most interesting and whose livability is truly refreshing. After a stay in this fine second home it will, indeed, be difficult to resume your daily activities in your first home. If you were to leave this leisure-living home you would surely miss the spaciousness of your living room, the efficiency of your work center, the pleasing layout of your master bedroom and all those glass sliding doors which mean you are usually but a step from out-of-doors.

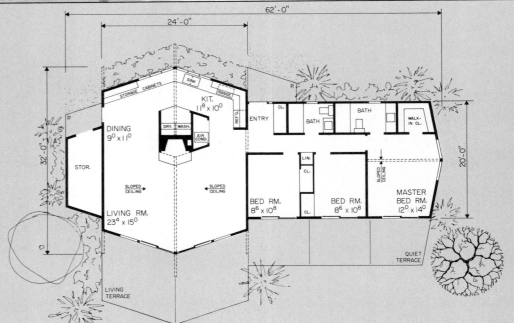

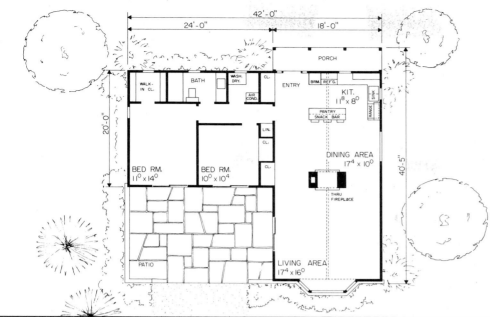

Design 32426
1,152 Sq. Ft.; 14,515 Cu. Ft.

● A touch of tradition pervades the environment around this L-shaped, frame, leisure-time home. The narrow horizontal sliding, the delicate window treatment and the prudent use of fieldstone, all help set the character. Inside, the floor plan offers wonderful livability. The huge living and dining areas are separated by an appealing thru-fireplace. Don't miss efficient kitchen. The island snack bar with pantry is a feature of the kitchen.

Design 31477
1,446 Sq. Ft.; 14,928 Cu. Ft.

● Who said you can't have a vacation home with French Provincial flair? The intriguing thought of having your own villa is certainly within the realm of distinct possibility. Call it what you like, this hip-roofed, brick veneer summer house has an inviting warmth you will love. Inside, there is space galore. Three bedrooms and two full baths in the sleeping wing. A huge living room with sloped ceiling. Storage cabinets in the kitchen plus a large storage area at the side entrance. List the other outstanding highlights.

Design 32458
1,406 Sq. Ft.; 14,108 Cu. Ft.

● The six-sided living unit of this design is highlighted by sloping ceilings and an abundance of glass to assure a glorious feeling of spaciousness. The homemaker's center is efficient and will be a delight in which to prepare and serve meals. It also highlights a large master bedroom and three bunk rooms. The bath facilities are compartmented and feature twin lavatories. To service the family's storage needs there are two walk-in closets, utility unit and a bulk storage unit.

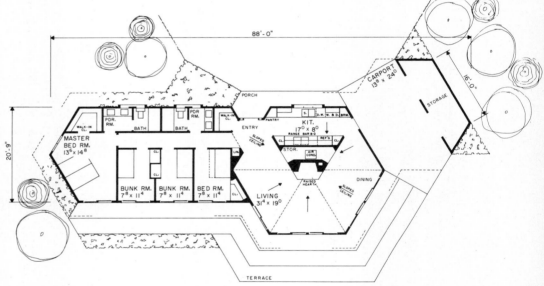

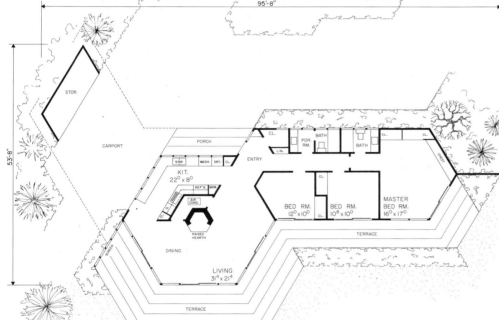

Design 31453
1,476 Sq. Ft.; 13,934 Cu. Ft.

● An exciting design, unusual in character, yet fun to live in. This frame home with its vertical siding and large glass areas has as its dramatic focal point a hexagonal living area with raised hearth fireplace which gives way to interesting angles. The large living area features sliding glass doors through which traffic may pass to the terrace stretching across the entire length of the house. The wide overhanging roof projects over the terrace and results in a large covered area outside the sliding doors of the master bedroom.

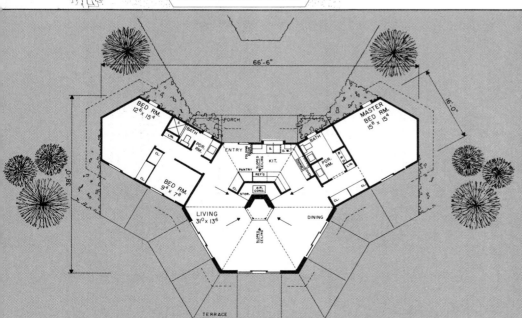

Design 32461
1,400 Sq. Ft.; 13,742 Cu. Ft.

● If you have the urge to make your vacation home one that has a distinctive flair of individuality, you should give consideration to the design illustrated here. Not only will you love the unique exterior appeal of your new home; but also, the exceptional living patterns offered by the interior. The basic living area is a hexagon. To this space conscious geometric shape is added the sleeping wings with baths. The center of the living area has as its focal point a dramatic raised hearth fireplace.

Design 32417
1,520 Sq. Ft.; 19,952 Cu. Ft.

● Have you ever seen a vacation home design that is anything quite like this one? Probably not. The picturesque exterior is dominated by a projecting gable with its wide overhanging roof acting as a dramatic sun visor for the wonderfully large glass area below. Effectively balancing this 20 foot center section are two 20 foot wings. Inside, and below the high sloping beamed ceiling, is the huge living area. In addition to the living/dining area, there is the spacious sunken lounge. This pleasant area has a built-in seating arrangement and a cozy fireplace.

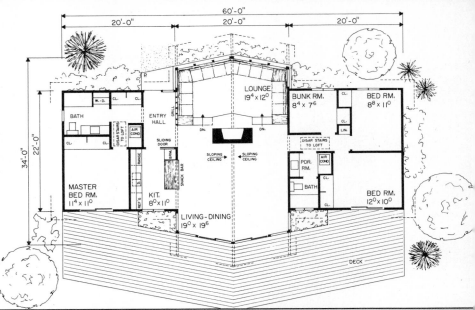

Design 31463
1,456 Sq. Ft.; 14,928 Cu. Ft.

● Your leisure-time home can represent a delightful departure from your year 'round home. Your choice of something new and different in the way of exterior design will be refreshing. Your selection of a unique, yet practical, floor plan will result in living patterns which will give your family a complete change of pace. This contemporary home highlights a spacious, center living core which is flanked by low-slung, flat-roofed sleeping wings. The over-all impact of the exterior is one of pure distinction. Inside, there is an atmosphere of cheerfulness fostered by the open planning, the walls of glass, the sloping ceilings and the fireplace. The children have their own sleeping area, while the parents have theirs.

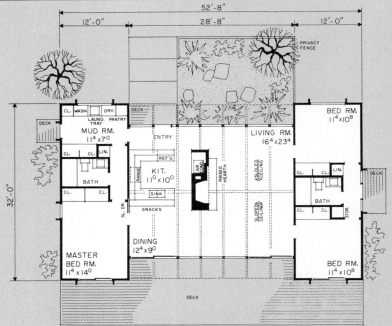

Design 31032
1,536 Sq. Ft.; 15,360 Cu. Ft.

● Here are contemporary living patterns which will surely be delightful. Particularly noteworthy is the planning of the bedrooms. Observe how the parents' and the children's bedrooms are situated at opposing ends of the house. A covered patio, which may be screened, enjoys exceptional privacy and its directly accessible from each sleeping area. Further, sliding glass doors open onto this outdoor area from the center living zone. The work center highlights a U-shaped kitchen, a built-in china cabinet and snack bar and the laundry equipment. A masonry wall provides screening for a front entrance court adjacent to the covered walk leading to the front door.

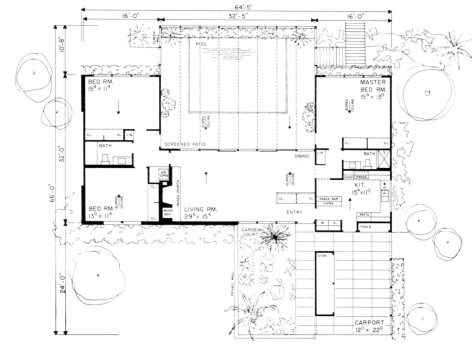

Design 32434 1,376 Sq. Ft.; 13,388 Cu. Ft.

● It should be easy to visualize the fun and frolic you, your family, your guests and your neighbors will have in this leisure-time home. The setting does not have to be near a bubbling brook, either. It can be almost any place where the pressures of urban life are far distant. The flat roof planes, the vertical brick piers, the massive chimney and the strategic glass areas are among the noteworthy elements of this design. Inside, there is space galore. The huge living-dining area flows down into the cozy, sunken lounge. The sleeping area of two bedrooms, a bath and good storage facilities is a zone by itself. The kitchen is efficient and has the bath and laundry equipment nearby. Imagine the spacious living area. It consists of a lounge, living room and dining room.

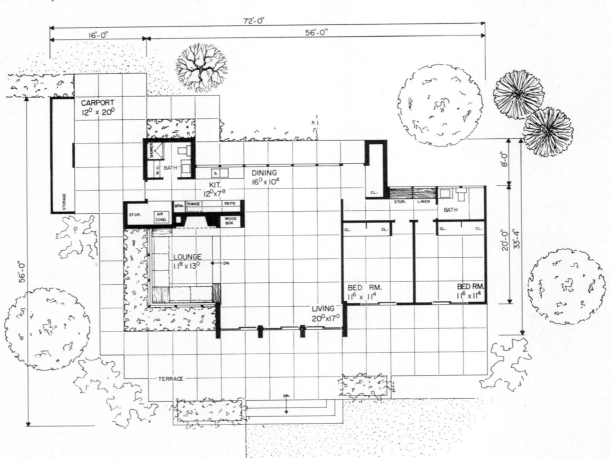

Design 32479   1,547 Sq. Ft.; 14,878 Cu. Ft.

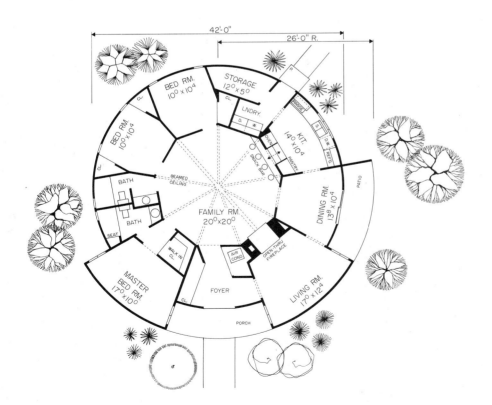

● Imagine living in a round house! Use it as a get-away or as a very unique year 'round home. Having the exterior shape so outstanding, creates an equally unique floor plan. All will marvel when they see how effectively this plan will function. All of the essential elements are available to serve you and your family. The centrally located family room is the focal point around which the various family functions and activities revolve. There is much to study and admire in this plan. For instance, the use of space is most efficient. Notice the strategic location of the kitchen. Don't miss the storage room and laundry. Observe the snack bar, the two-way fireplace, the separate dining room and the two full baths. Fixed glass windows at the beamed ceiling provides natural light from above for the family room.

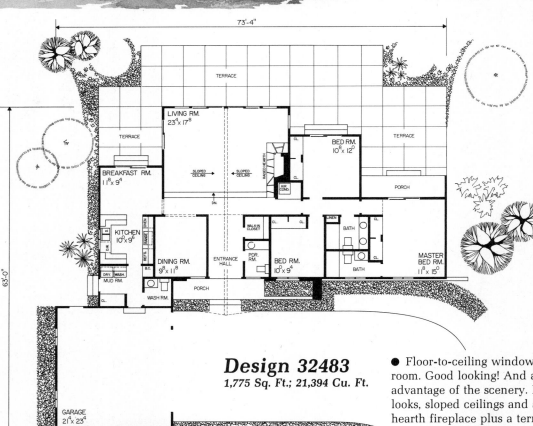

Design 32483
1,775 Sq. Ft.; 21,394 Cu. Ft.

● Floor-to-ceiling windows in the living room. Good looking! And a way to take advantage of the scenery. For more good looks, sloped ceilings and a raised hearth fireplace plus a terrace that runs the length of the house. Dining can be enjoyed in the formal dining room and the efficient U-shaped kitchen's separate breakfast nook. A laundry/mud room that allows immediate clean-up after a day spent fishing or on the beach. Three bedrooms! Including one with a private bath. A great vacation home. Good for living all-year-round.

NOTE: Interested in reviewing additional vacation home designs? If so, the plan book, *223 Vacation Homes*, is available at $4.25 per copy. It contains 176 pages of which 96 are in full color. A wide variety of chalets, A-frames and cottages from 480 to 2928 square feet. Send your remittance to Home Planners, Inc., Dept. CV, 23761 Research Drive, Farmington Hills, Michigan 48024.

One-Story Multi-Family Living

Design 32018
1,267 Sq. Ft. - Each Unit
19,478 Cu. Ft. - Each Unit

● Here is a series of multi-family homes which feature on one floor, complete family living potential. In some cases the basic structure is a one-story duplex, in others, it is a two-story building. With today's high construction costs, you may find that this type of living unit makes economic sense. Friends or families may want to consider sharing the expense of such a building program. Or, you may wish to build such a home yourself and rent one of the units while occupying the other. In any event, your family's living patterns will be as convenient and efficient as those enjoyed by occupants of a single family, ranch home. The design on this page is a refreshing contemporary duplex. The raised brick planter and privacy wall are distinctive features. Sliding glass doors open onto the dining terrace. (The terraces are on opposite sides of the house for maximum privacy.)

Design 32016
1,008 Sq. Ft. - Each Unit
11,390 Cu. Ft. - Each Unit

Design 32019
960 Sq. Ft. - Each Unit
10,220 Cu. Ft. - Each Unit

● Three duplexes! Here are three two-family houses, each with two living units all on one floor. Ideal for those contemplating the ownership of income property. A study of each of the floor plans reveals differing livability patterns. Whatever your choice, you'll find maximum use of space which will assure convenient living and economical construction. Surely, a fine return on your investment dollar.

Design 32039
1,139 Sq. Ft. - Each Unit
12,481 Cu. Ft. - Each Unit

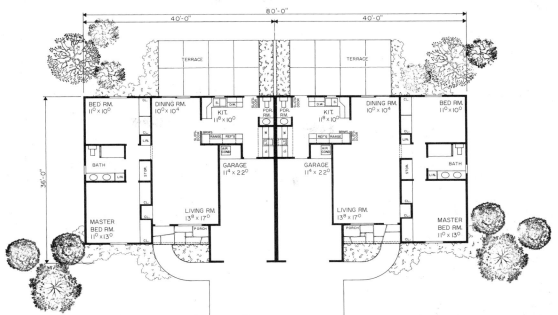

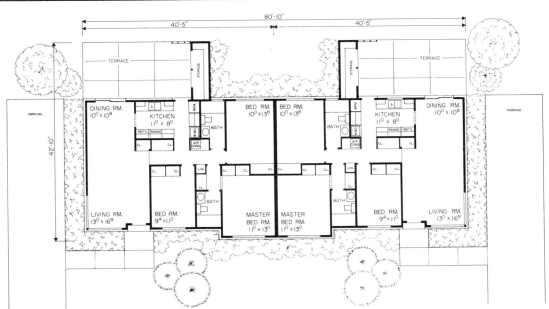

An Efficient Duplex Plan...

Design 32021
956 Sq. Ft. - Each Unit
11,983 Cu. Ft. - Each Unit

Design 32022
956 Sq. Ft. - Each Unit
11,286 Cu. Ft. - Each Unit

Design 32023
956 Sq. Ft. - Each Unit
11,036 Cu. Ft. - Each Unit

...Having a Choice of Three Exteriors

● These three beautifully styled duplexes each go with this one efficient floor plan. The three exteriors are very different in style: Design 32021 is a traditional with coach lamps and cupola, Design 32023 is a contemporary with vertical paned windows and Design 32022 is an elegant French design. After you choose the exterior, imagine the livability of the floor plan. Two bedrooms with a nearby bathroom. The kitchen has sliding glass doors to the terrace. The L-shaped living/dining room is where your guests will be greeted. This plan includes details for an optional non-basement. When the non-basement plan is used, additional storage space is added for your convenience. This home meets the three standards of being attractive, efficient and very economical.

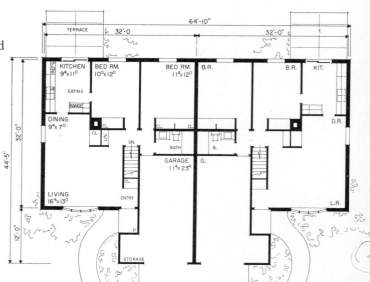

... Three More Popular Titles —

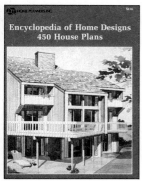

23 320 Pages, $8.95

24 128 Pages, $2.95

25 192 Pages, $3.50

450 HOUSE PLANS - An encyclopedia of home designs for those who wish to review and study perhaps the largest selection of designs available in a single volume. This book features sections on Heritage Houses from our architectural past along with Trend Houses, one, 1½ and two-story homes, multi-levels and vacation homes. Of particular interest are designs with optional exteriors and floor plans. Also, homes for country-estate living.

152 HOUSE PLANS - An appealing use of second color tones effectively compliments the wide variety of exterior styles and practical floor plans. There are one, 1½, two-story and multi-level house plans. From 936 to 4509 square feet. For small, medium and large families. Included are houses with second floor lounges, secluded master suites, country kitchens, gathering rooms, exposed lower levels, indoor and outdoor balconies and energy-oriented solariums and greenhouses.

255 HOME DESIGNS FOR FAMILY LIVING - This exciting two-color plan book has over 700 illustrations of contemporary and favorite traditional exteriors. Tudor, French, Spanish and Early Colonial are among the popular styles. In addition to the plans which cater to a wide variety of family living patterns and budgets, there are special feature sections on: Vacation Homes, Earth-Sheltered Designs, Sun-Oriented Living and Shared Livability.

THE PLAN BOOKS

... are a most valuable tool for anyone planning to build a new home. A study of the hundreds of delightfully designed exteriors and the practical, efficient floor plans will be a great learning and fun-oriented family experience. You will be able to select your preferred styling from among Early American, Tudor, French, Spanish and Contemporary adaptations. Your ideas about floor planning and interior livability will expand. And, of course, after you have selected an appealing home design that satisfies your long list of living requirements, you can order the blueprints for further study of your favorite design in greater detail. Surely the hours spent studying the portfolio of Home Planners' designs will be both enjoyable and rewarding ones.

Kindly note: For detailed information about the complete home planning package, see pages 241 through 246.

HOME PLANNERS, INC.
Dept. BK, 23761 Research Drive
Farmington Hills, Michigan 48024

Phone Toll Free:
1-800-521-6797

PLAN BOOK ORDER FORM

Please mail me the following:

<u>THE DESIGN CATEGORY SERIES</u> - *A great series of books specially edited by design type and size. Each book features interesting sections to further enhance the study of design styles, sizes and house types. A fine addition to the home or office library. Complete collection - over 1275 designs.*

Unit Price

1. _____ 400 1½ & Two-Story Home Plans .... $5.95 $_____
2. _____ 210 One-Story - Over 2,000 sq. ft. .... $3.95 $_____
3. _____ 350 One-Story - Under 2,000 sq. ft. .... $4.95 $_____
4. _____ 205 Multi-Level Home Plans .... $3.95 $_____
5. _____ 223 Vacation Homes .... $4.25 $_____

<u>OTHER CURRENT TITLES</u> - *The interesting series of plan books listed below have been edited to appeal to various style preferences and budget considerations. The majority of the designs highlighted in these books also may be found in the Design Category Series.*

The Budget Series-
6. _____ 175 Low Budget Homes.................................... $2.75 $_____
7. _____ 165 Affordable Home Plans .... $2.95 $_____
8. _____ 142 Home Designs for Expanded Bldg. Budgets .... $2.75 $_____
9. _____ 110 Home Plans .... $2.75 $_____

The Exterior Style Series-
10. _____ 120 Early American Home Plans .... $2.95 $_____
11. _____ 125 Contemporary Home Plans .... $2.95 $_____
12. _____ 135 English Tudor Homes .... $2.95 $_____
13. _____ 136 Spanish & Western Home Designs *(July '85)* .... $2.95 $_____
14. _____ 130 Distinctive Home Designs........................... $2.75 $_____
15. _____ 144 Home Designs For All Americans .... $2.95 $_____
16. _____ 112 Traditional & Contemporary Family Homes .... $2.95 $_____
17. _____ 102 Home Plans .... $2.75 $_____

Three Great Books In Full Color-
18. _____ 116 Traditional & Contemporary Plans .... $5.95 $_____
19. _____ 122 Home Designs .... $5.95 $_____
20. _____ 114 Trend Homes .... $5.95 $_____

Two Books Of Most Popular Designs-
21. _____ 166 Most Popular Homes .... $2.95 $_____
22. _____ 172 Most Popular Homes .... $2.95 $_____

Encyclopedia Of Home Designs-
23. _____ 450 House Plans.. $8.95 $_____

Two More Outstanding New Titles-
24. _____ 152 House Plans.. $2.95 $_____
25. _____ 255 Home Designs for Family Living .... $3.50 $_____

MAIL TODAY   **SATISFACTION GUARANTEED!**
         Your order will be processed
         and shipped within 48 hours

Sub Total $_____
Michigan Residents
kindly add 4% sales tax $_____
TOTAL-Check enclosed $_____

Name _____
Address _____
City _____ State _____ Zip _____

In Canada Mail To: Home Planners, Inc. 20 Cedar St. N., Kitchner, Ontario N2H 2W8

CV3

Design 32038

1,699 Sq. Ft. - Each Unit
Excluding Atrium
18,105 Cu. Ft.

● Contemporary styling at its best! This is surely an outstanding duplex home. Note the fine exterior features: the overhanging hip roof, the privacy wall with garden court and the projecting garage to the front and bedroom wing to the rear to form a natural privacy wall. The interior livability is also outstanding. The large living room has a sloped ceiling, built-in book shelves and cabinets and a centered fireplace with wood box on one side and storage on the other. Efficient L-shaped kitchen easily serves the dining room, family room and the snack bar on the terrace. The bedroom wing houses the three bedrooms and two baths. Notice the sliding glass doors in the dining room leading to the rear terrace and another garden court. This area is accessible from each of the bedrooms yet has maximum privacy from outsiders with the gate closing the area off. The atrium is an interesting focal point. It is 115 square feet. There are many extras featured in the garage.

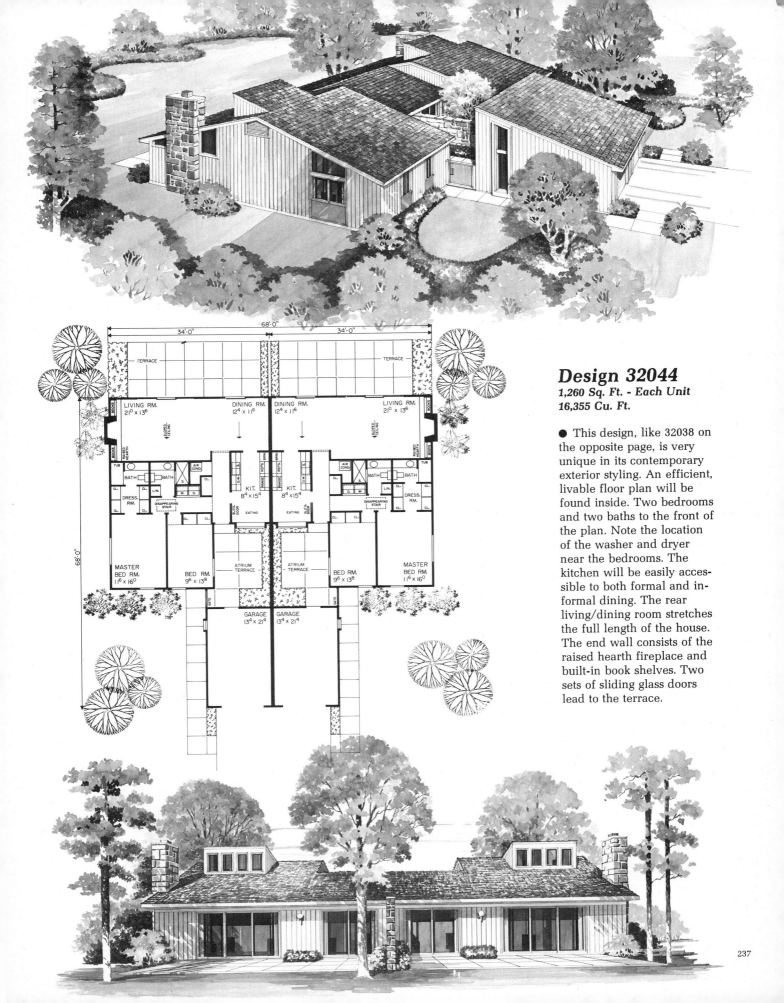

Design 32044
1,260 Sq. Ft. - Each Unit
16,355 Cu. Ft.

● This design, like 32038 on the opposite page, is very unique in its contemporary exterior styling. An efficient, livable floor plan will be found inside. Two bedrooms and two baths to the front of the plan. Note the location of the washer and dryer near the bedrooms. The kitchen will be easily accessible to both formal and informal dining. The rear living/dining room stretches the full length of the house. The end wall consists of the raised hearth fireplace and built-in book shelves. Two sets of sliding glass doors lead to the terrace.

237

Design 32828
First Floor: 817 Sq. Ft. - Living Area; 261 Sq. Ft. - Foyer & Laundry
Second Floor: 852 Sq. Ft. - Living Area; 214 Sq. Ft. - Foyer & Storage; 34,690 Cu. Ft.

TWO COUPLES/SINGLES RESIDENCE

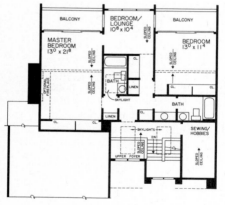

CONVERTIBLE ONE-FAMILY RESIDENCE

● This contemporary home has been designed as a two-couples/singles residence. A home of this type could be bought jointly by two couples or one couple could buy the entire home and rent out one of the units. Complete livability is offered on each floor of this two-story. Each floor has a living room, dining room, interior kitchen, bedroom and bath. At a later date this home could be converted into a one-family residence. The second floor unit would now be a bedroom area.

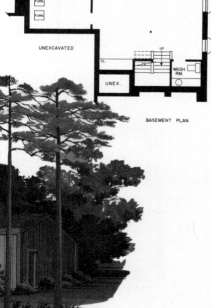

BASEMENT PLAN

238

Shared Expense – Shared Livability

Design 32869
1,986 Sq. Ft.; 48,455 Cu. Ft.

● This traditional one-story design offers the economical benefits of shared living space without sacrificing privacy. The common area of this design is centrally located between the two private, sleeping wings. The common area, 680 square feet, is made up of the great room, dining room and kitchen. Sloping the ceiling in this area creates an open feeling as will the sliding glass doors on each side of the fireplace. These doors lead to a large covered porch with skylights above. Separate outdoor entrances lead to each of the sleeping wings. Two bedrooms, dressing area, full bath and space for an optional kitchenette occupy 653 square feet in each wing. Additional space will be found in the basement which is the full size of the common area. Don't miss the covered porch and garage with additional storage space.

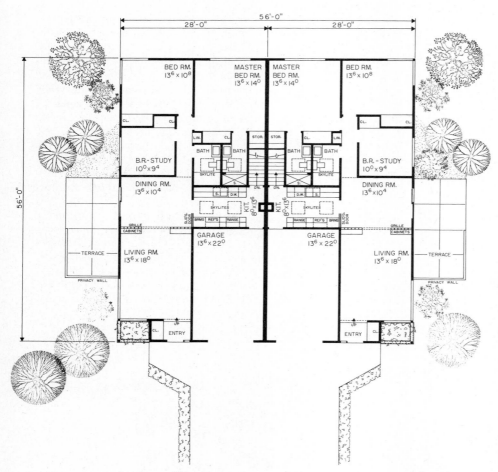

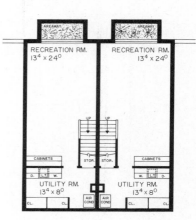

Design 32017
1,233 Sq. Ft. - Each Unit
17,599 Cu. Ft. - Each Unit

● This modern one-story duplex is surely eye-catching. This house is a perfect square which will cut down building costs. Note how the garage is part of the main house. Also for savings, the plumbing facilities (two baths and kitchen in both units) have been grouped together. The living room and dining room are partially separated by handsome grill work with cabinets below. A sliding door closes the kitchen off from the dining room. Three bedrooms or two plus a study will serve the family. The side terrace, accessible by sliding glass doors in the dining room, is sheltered from the street by a privacy wall. The basement may be developed as a recreation area as shown at the right.

NOTE: Additional multi-family designs can be found in *Multi-Family Living*. It features 72 designs and is available at $3.50 per copy. Included are duplexes, fourplexes, townhouses, attached apartments, shared-livability, common areas and accommodations for the live-in relative. Remittance should be sent to Home Planners, Inc., Dept. CV, 23761 Research Drive, Farmington Hills, Michigan 48024.

ALL the "TOOLS" you and your builder need...

... to, first select an exterior and a floor plan for your new house that satisfy your tastes and your family's living patterns ...

... then, to review the blueprints in great detail and obtain a construction cost figure ... also, to price out the structural materials required to build ... and, finally, to review and decide upon the specifications to which your home is to be built. Truly, an invaluable set of "tools" to launch your home planning and building programs.

1. THE PLAN BOOKS
Home Planners' unique Design Category Series makes it easy to look at and study only the types of designs for which you and your family have an interest. Each of five plan books features a specific type of home, namely: 1½ and 2-Story, One-Story Over 2000 Sq. Ft., One-Story Under 2000 Sq. Ft., Multi-Levels and Vacation Homes. In addition to the convenient Design Category Series, there is an impressive selection of other current titles. While the home plans featured in these books are also to be found in the Design Category Series, they, too, are edited for those with special tastes and requirements. Your family will spend many enjoyable hours reviewing the delightfully designed exteriors and the practical floor plans. Surely your home or office library should include a selection of these popular plan books. Your complete satisfaction is guaranteed.

2. THE CONSTRUCTION BLUEPRINTS
There are blueprints available for each of the designs published in Home Planners' current plan books. Depending upon the size, the style and the type of home, each set of blueprints consists of from five to ten large sheets. Only by studying the blueprints is it possible to give complete and final consideration to the proper selection of a design for your next home. The blueprints provide the opportunity for all family members to familiarize themselves with the features of all exterior elevations, interior elevations and details, all dimensions, special built-in features and effects. They also provide a full understanding of the materials to be used and/or selected. The low-cost of our blueprints makes it possible and indeed, practical, to study in detail a number of different sets of blueprints before deciding upon which design to build.

3. THE MATERIAL LIST
A list of materials is an integral part of the plan package. It comprises the last sheet of each set of blueprints and serves as a handy reference during the period of construction. Of course, at the pricing and the material ordering stages, it is indispensable.

4. THE SPECIFICATION OUTLINE
Each order for blueprints is accompanied by one Specification Outline. You and your builder will find this a time-saving tool when deciding upon your own individual specifications. An important reference document should you wish to write your own specifications.

THE PLAN BOOKS
The Design Category Series...

...A great selection of five plan books specially edited for ease in studying specific design types. Features most of the house plans shown in other current titles. The five book Complete Collection guarantees many enjoyable hours of happy house hunting. A fine set of books for the home or office reference library.

1. **400 1½ and TWO-STORY HOME PLANS** - Those interested in studying a wide variety of 1½ and two-story exteriors and floor plans need look no further. New England Gambrels, Salt Boxes, Tudor, French Mansards, Georgians, Southern Colonials, Cape Cods, Virginia Tidewater, Farmhouses and Contemporary exteriors are featured. Family living floor plans with two to six bedrooms.

2. **210 ONE-STORY HOME PLANS** - Over 2,000 Square Feet - Designs for those who prefer one-story living and all the convenience that goes with it. A selection of homes with varying exterior styles housing practical and efficient family living floor plans. Gathering rooms, family rooms, formal and informal dining areas.

3. **350 ONE-STORY HOME PLANS** - Under 2,000 Square Feet - A wide selection of one-story homes for the modest building budgets. Delightful Traditional exteriors as well as exciting Contemporaries. Fine functioning floor plans for both the small and large family. Plans with optional elevations.

4. **205 MULTI-LEVEL HOME PLANS** - For those who wish to experience new dimensions in total livability. This fine collection includes split foyer bi-levels and tri-levels for flat and sloping sites. Also, homes with exposed lower levels.

5. **223 VACATION HOMES** - A popularly acclaimed vacation and leisure-living book of exteriors and floor plans. A-Frames, Chalets, Hexagons and other interesting shapes with decks, balconies and terraces. 96 exciting full color pages.

1 320 Pages, $5.95

2 192 Pages, $3.95

3 256 Pages, $4.95

4 192 Pages, $3.95

5 176 Pages, $4.25

The Exterior Style Series...

...Delightfully edited for those who wish to review home plans of their favorite exterior styling. Ideal for those who want to compare the unique appeal of various pleasing facades.

10. **120 EARLY AMERICAN PLANS** - This unique plan book is devoted exclusively to Early American architectural interpretations adapted for today's living patterns. Exquisitely detailed exteriors retain all of the charm of a proud heritage.

11. **125 CONTEMPORARY HOME PLANS** - An exciting book featuring a wide variety of home designs for the 1980's and far beyond. The exteriors are refreshing with their practical and progressive "new look".

12. **135 ENGLISH TUDOR HOMES** - and other Popular Family Plans is a favorite of many. The current popularity of the English Tudor home design is phenomenal and this book is loaded with Tudors for all budgets.

13. **136 SPANISH & WESTERN HOME DESIGNS** - Stucco exteriors, tile roofs, courtyards and rambling ranches are characteristics which make this design selection distinctive. These sun-country designs highlight indoor-outdoor relationships. Their appeal is not limited to the Southwest region of our country.

14. **130 DISTINCTIVE HOME DESIGNS** - Commencing with the Early American Homes and continuing through the Tudor, French, Spanish and Contemporary, this book will be enjoyed by all. The pleasing exteriors and exciting floor plans assure hours of rewarding home planning.

15. **144 HOME DESIGNS FOR ALL AMERICANS** - Here, favorite house styles match geographical regions. The New England section has its Gambrels; South, its stately columns; West, its Spanish charm; Northwest, its dramatic contemporary design; Mid-Atlantic, its historic front porch, etc.

16. **112 TRADITIONAL and CONTEMPORARY FAMILY HOMES** - A delightful collection of designs for varying tastes. All sizes and types of designs for family living. Over 300 exterior and floor plan illustrations.

17. **102 HOME PLANS** - An excellent selection of home designs featuring a wide variety of exterior styles. There are Early American, Tudor, Spanish, French and Contemporary facades. A special 16 page section in full color.

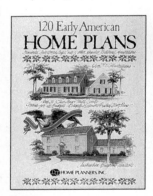
10 112 Pages, $2.95

11 112 Pages, $2.95

12 104 Pages, $2.95

14 112 Pages, $2.75

15 128 Pages, $2.95

16 96 Pages, $2.75

The Budget Series . . .

. . . Construction costs are influenced to a significant extent by the size of the house. The houses and plans in this series have been edited according to square footage ranges. Each book highlights a wide variety of design styles and types.

6 96 Pages, $2.75

7 112 Pages, $2.95

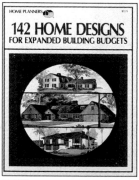
8 112 Pages, $2.75

9 112 Pages, $2.75

175 LOW BUDGET HOMES - The house designs in this book average 1505 square feet. From 1165 square foot average for the one-story houses to an 1812 average for two-stories. 1½-Story and tri-level averages fall within this range, too. Many designs have expansive potential where sleeping areas may be finished off later. Wide selection of styles. Two to five bedrooms. Family rooms, extra baths, formal and informal dining rooms. This book is a must for those with a restricted building budget.

165 AFFORDABLE HOME PLANS - Designs averaging 2052 square feet are featured in this collection of houses. They range from a 1581 square foot average for the one-stories, to 2261 for the two-story homes, to 2381 for the tri-levels. Tudor, French, Early American, Spanish and Contemporary exteriors are featured throughout the book. Efficient, family living floor plans. This wide selection of houses and plans will fit the medium budget. Basement and non-basement designs.

142 HOME DESIGNS FOR EXPANDED BUILDING BUDGETS - This selection of designs highlight houses with an average square footage of 2551. One-story plans average 2069; two-stories, 2735; multi-levels, 2825. As the family's size and income grows so does its need for, and ability to finance, a larger home grow. A fine group of designs for all exterior style tastes and livability requirements. Spacious homes featuring raised hearth fireplaces, beamed ceilings, open planning and efficient kitchens.

110 HOME PLANS FOR VARYING BUILDING BUDGETS - Edited in appealing two-color featuring designs for all budgets. One, 1½, two-story and multi-levels. Colonial, Tudor, Spanish, French and Contemporary exteriors, among the most popular, are featured. Special section of energy-oriented designs with solariums, atriums, skylights, collectors, etc. Trend houses and history house designs. Houses designed for flat and hillside sites. Exposed lower levels are also available.

Three Great Books in Full Color . . .

For Plan Book Order Form Kindly turn to page 235.

13 96 Pages, $2.95

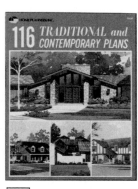
18 96 Pages, $5.95

19 96 Pages, $5.95

20 96 Pages, $5.95

116 TRADITIONAL and CONTEMPORARY PLANS - A beautifully illustrated home plan book in complete, full color. One, 1½, two-story and split-level designs featured in all of the most popular exterior styles.

122 HOME DESIGNS - This book has delightfully dramatic full color throughout. More than 120 eye-pleasing, colored illustrations. Tudor, French, Spanish, Early American and Contemporary exteriors.

114 TREND HOMES - Heritage Houses, Energy Designs, Family Plans - these, along with Vacation Homes, are in this exciting, new plan book in full color. A potpourri of designs cater to a variety of tastes.

17 96 Pages, $2.75

Popular Designs . . .

166 MOST POPULAR HOMES - A book of best-selling house plans containing over 400 illustrations. Houses range in size from 1,050 to 5,308 square feet. Tudor, Early American, Spanish and French exteriors plus Contemporary elevations and floor plans.

172 MOST POPULAR HOMES - The second book in this series. These designs are selected from Home Planners esteemed portfolio of over 1400 different home plans. An excellent book to gauge the tastes in styles and preferences in floor plans from the past readers of Home Planners' books.

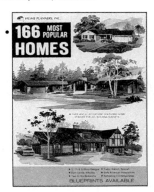
21 112 Pages, $2.95

22 128 Pages, $2.95

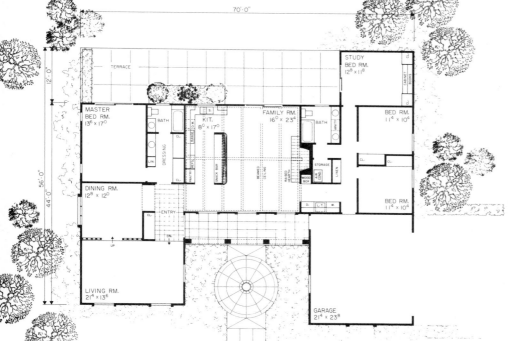

Design 32236
2,307 Sq. Ft.; 28,800 Cu. Ft.

● Living in this Spanish adaptation will truly be fun for the whole family. It will matter very little whether the backdrop matches the mountains below. A family's flair for distinction will be satisfied by this picturesque exterior, while its requirements for everyday living will be gloriously catered to. The hub of the plan will be the kitchen-family room area. The beamed ceiling and raised hearth fireplace will contribute to the cozy, informal atmosphere. The separate dining room and the sunken living room function together formally.

251

Design 32283 1,559 Sq. Ft. - First Floor
1,404 Sq. Ft. - Second Floor; 48,606 Cu. Ft.

● Reminiscent of the stately character of Federal architecture during an earlier period in our history, this two-story is replete with exquisite detailing.

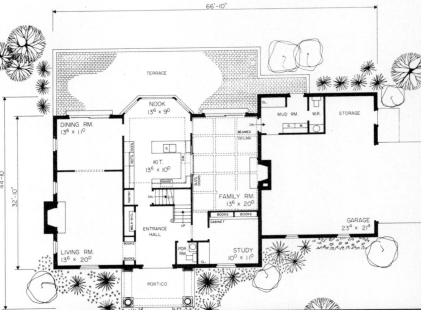

Design 32520 1,419 Sq. Ft. - First Floor
1,040 Sq. Ft. - Second Floor; 39,370 Cu. Ft.

● From Tidewater Virginia comes this historic adaptation, a positive reminder of the charm of Early American architecture. Note how the center entrance gives birth to fine traffic circulation. Corner fireplaces in both the family room and living room are surely delightful.

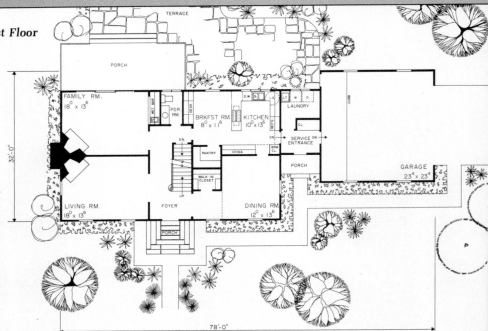

Design 32131   1,214 Sq. Ft. - First Floor; 1,097 Sq. Ft. - Second Floor; 30,743 Cu. Ft.

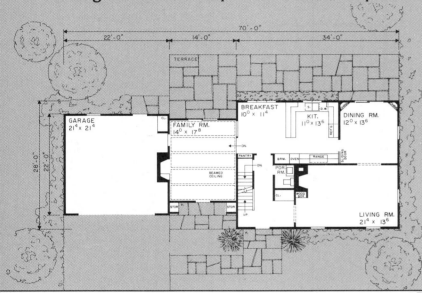

● The Gambrel-roof home is often the very embodiment of charm from the Early Colonial Period in American architectural history. Fine proportion and excellent detailing were the hallmarks of the era. Notice the window and door treatment; the narrow siding and the corner boards; the massive chimney. An appealing highlight is the covered service porch. This covered entrance with its adjacent twin storage units provides the family with access to the front yard.

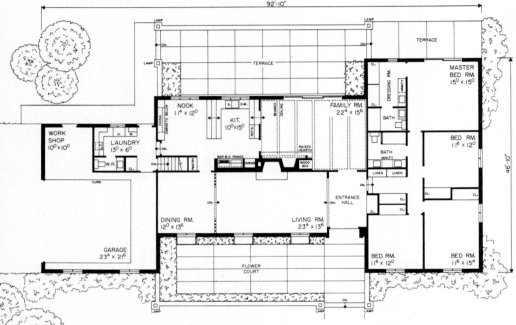

Design 32220
2,646 Sq. Ft.; 46,880 Cu. Ft.

● The gracious formality of this home is reminiscent of a popularly accepted French styling. The hip-roof, the brick quoins, the cornice details, the arched window heads, the distinctive shutters, the recessed double front doors, the massive center chimney and the delightful flower court are all features which set the dramatic appeal of this home. This floor plan is a favorite of many. The four bedroom, two bath sleeping wing is a zone by itself. Further, the formal living and dining rooms are ideally located. They function well together for entertaining and look out upon the pleasant front flower court.

Design 31754
2,080 Sq. Ft.; 21,426 Cu. Ft.

● Boasting a traditional Western flavor, this rugged U-shaped ranch home has all the features to assure grand living. The private front flower court, inside the high brick wall, creates a delightfully dramatic atmosphere which carries inside. The floor plan is positively unique and exceptionally livable. Wonderfully zoned, the bedrooms enjoy their full measure of privacy. The formal living and dining rooms function together in a most pleasing fashion. The areas of the laundry, kitchen, informal eating and family room fit together in such a manner as to guarantee efficient living patterns.

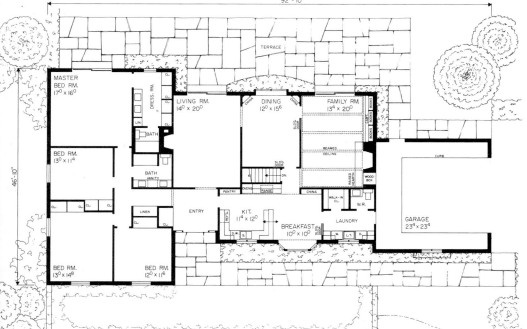

Design 32181
2,612 Sq. Ft.; 45,230 Cu. Ft.

● It is hard to imagine a home with any more eye-appeal than this one. It is the complete picture of charm. The interior is no less outstanding. Sliding glass doors permit the large master bedroom, the quiet living room and the all-purpose family room to function directly with the outdoors. The two fireplaces, the built-in china cabinets, the book shelves, the complete laundry and the kitchen pass-thru to breakfast room are extra features. Count the closets. Don't miss those in the beamed ceilinged family room. Although the illustration of this home shows natural quarried stone, you may wish to substitute brick or even siding.

255

Design 32128 1,152 Sq. Ft. - First Floor
896 Sq. Ft. - Second Floor; 30,707 Cu. Ft.

● Here is proof that your restricted building budget can return to you wonderfully pleasing design and loads of livability. This is an English Tudor adaptation that will surely become your subdivision's favorite facade. Its mark of individuality is obvious to one and all. The interior is economically compact without being at all cramped. Every feature is present to guarantee the complete livability sought by today's active family. There are the four bedrooms, the 2½ baths, the two eating areas, the quiet living room, the well-planned kitchen and the attached two-car garage. Don't miss the two fireplaces and the huge storage area.

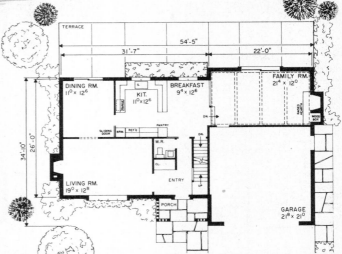

Design 31791 1,157 Sq. Ft. - First Floor
875 Sq. Ft. - Second Floor; 27,790 Cu. Ft.

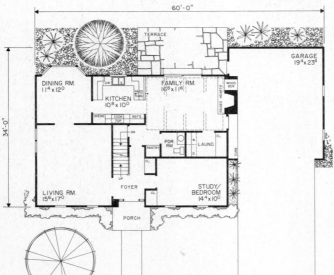

● Wherever you build this house an aura of Cape Cod is sure to unfold. The symmetry is pleasing, indeed. The authentic center entrance seems to project a beckoning call. Although small and cozy in appearance, the floor plan is long on livability. In addition to the three bedrooms upstairs there is another on the first floor. A pass-thru between the kitchen and family room is a handy feature.

NOTE: Outstanding selection of popular designs is available at $2.95 per copy. Order *166 Most Popular Homes* from Home Planners, Inc., Dept. CV, 23761 Research Drive, Farmington Hills, Michigan 48024.